Industrial chemistry
case studies

ROYAL SOCIETY OF CHEMISTRY

Industrial chemistry
case studies

Compiled by Ted Lister

Edited by Colin Osborne and John Johnston

Designed by Imogen Bertin and Sara Roberts

Published by the Education Division, The Royal Society of Chemistry

Printed by The Royal Society of Chemistry

For further information on other educational activities undertaken by the Royal Society of Chemistry write to:

The Education Department
The Royal Society of Chemistry
Burlington House
Piccadilly
London W1V OBN

ISBN 0 85404 925 8

British Library Cataloguing in Data.
A catalogue for this book is available from the British Library.

RS•C

Contents

RS•C

Introduction

This publication presents material for chemistry teachers on a variety of aspects of the UK chemical industry (including the steel industry) in the late 1990s. In particular it aims to cover the economic dimension of the chemical industry, some environmental concerns and constraints and the issue of scaling up processes from the laboratory bench to full scale production *via* pilot plant.

It is designed to be read by teachers to update themselves personally and also to be used by post-16 students (A-level, AS, Scottish Higher and GNVQ and GSVQ). To this end there are questions (with answers for the teacher) within the text. The questions deal with chemical principles which will be familiar to students at this level but which are set in relatively unfamiliar contexts taken from industrial chemistry. A case study approach has been adopted with studies drawn from a number of areas of the chemical industry. Many of the studies deal with current projects and, from time to time, issues of commercial sensitivity have been encountered which have meant that it has not been possible to go into as much detail as might have been wished.

Nomenclature and units

Chemistry students in schools and colleges are used to systematic chemical nomenclature based on the IUPAC rules. Much of the chemical industry uses trivial names; indeed in many cases, systematic names for complex chemicals are somewhat cumbersome. It has therefore been a matter of judgement as to which name to use for a particular chemical in a particular situation. Sometimes the nomenclature adopted by industry has been used, for example the ethylene (not ethene) pipeline. In other cases, the systematic name has been used in the study, even though industry will use a trivial one (for example phenylmethanol, rather than benzyl alcohol). Where appropriate, both names have been given; the main aim has been to avoid confusion and ambiguity.

Many of the same sorts of considerations occur with units. In general, units have been converted into those used by students in UK schools and colleges.

RS•C

Acknowledgements

This publication was made possible with the help of many people who gave freely (in both senses of the word) of their time. Particular thanks go to the following people and their organisations.

Pauline Fawcett	British Nuclear Fuels plc
Paul Brooks, Joe Gallagher, Colin Green, Richard Leonard, Barry Prater, David Shillaker, Kevin Whitty and Graham Williams-Buckley	British Steel plc
Brian Moffatt	Shell Chemicals
Mike Fedouloff and Sue James	SmithKline Beecham Pharmaceuticals
Elspeth Gray	Abbott Laboratories
Simon Baker, Robert Joseph, David Knee, Stephen Maund and Keith Parry	Zeneca Agrochemicals
Narinder Kehal and Neil Steptoe	Raychem
Ken Hepburn	Intertek Testing Services
Steve Wallace	British Gas
Robert Cutler and Paul Board	Robertson Laboratories
David Llewellyn	Cambrian Stone
Malcolm Braithwaite	Rexchem organics
Libby Steele	The Association of the British Pharmaceutical Industry

Thanks also go to David Moore of St Edward's School, Oxford and Samina Khan of Sutton Coldfield College who read and commented on drafts of parts of the manuscript and to John Johnston and Colin Osborne of The Royal Society of Chemistry who read the proofs.

RS•C

The UK chemical industry in the 1990s – a brief overview

The UK chemical industry is the nation's fifth largest industrial sector (after food, drink and tobacco; paper, printing and publishing; electrical engineering; and transport equipment). It is estimated that the industry consists of some 3600 companies making almost 100,000 different substances. Sales in 1995 were just under £40 billion. Over half these sales are exports, making the industry the UK's largest export earner. On a world scale, the UK chemical industry is the fifth largest national chemical industry, after those of the US, Japan, Germany and France. The UK chemical industry is growing at a rate of 3.7% per year - over twice the rate of growth of UK manufacturing industry as a whole. Almost a quarter of a million people are employed directly by the industry in the UK and many more are employed indirectly. *Figure 1* shows how these jobs are distributed around the country.

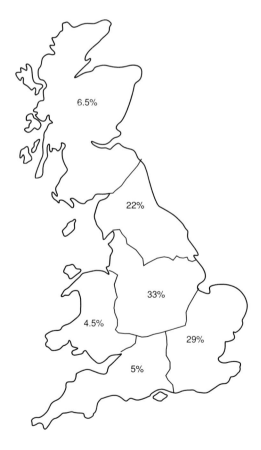

**Figure 1 Chemical industry employment by region, % shares of GB
(courtesy of Chemical Industries Association, CIA)**

In 1995, the chemical industry made just over £1 billion of capital investment. *Figure 2* shows how the chemical industry is broken down into different sectors while Table 1 shows that the average household spends £25 weekly on a variety of chemical products. These include direct purchases of chemicals, such as household bleach or petrol, and indirect ones, such as a pair of shoes whose leather has been treated by a variety of chemicals.

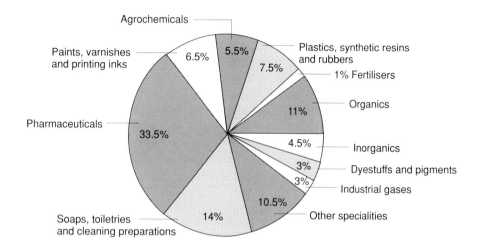

Figure 2 UK chemical industry sector shares of gross value added, 1993 (courtesy of CIA)

	£
Food, drink, *etc*	6.00
Clothing, footwear	2.10
Health, personal care	4.75
Housing, household goods	5.00
Travel, motoring, telecoms	3.50
Entertainment	3.65
Total	**25.00**

Table 1 Through the products and services it consumes, each UK household either directly or indirectly spends an average of £25 per week on chemicals (courtesy of CIA)

RS•C

Individual companies

Almost all of the UK chemical industry is privately owned rather than state owned. The last major publicly owned concern was BP, which was sold in 1987. Much of the industry consists of large multinational organisations, such as GlaxoWellcome, formed in 1995, which is the world's largest pharmaceutical company. On the other hand, half of UK chemical companies employ fewer than 100 staff.

Location

Traditionally, heavy chemical manufacturing has been clustered around ports. These allow for easy export and import of products and raw materials. This is still the case to some extent, particularly for chemicals made in large tonnages, and there are still concentrations of the industry in the North West of England, the North East (Teesside) and Humberside as well as in Scotland, Bristol, Southampton and the Thames. Other significant locations are near the sites of raw materials such as around the Northwich salt deposits or St Austell's clay. However, the siting of plant which manufacture small volumes of high value speciality chemicals and the locations of corporate headquarters and research and development facilities are less affected by this type of consideration. Many of these are located in the South East of England.

The structure of the chemical industry

What the chemical industry essentially does is to start with raw materials in large volume and process them in several stages so that the products of each stage have a higher value than the starting materials. So, as they are increasingly processed, the chemicals pass along a continuum from low value, high tonnage to low tonnage, high value. A number of terms are associated with the various stages.

▼ **Raw materials** are the basic starting materials of chemical processing and are extracted from natural sources – they include crude oil, coal, air, mineral-bearing rocks and limestone. However, it is worth appreciating that the waste materials from one plant or process can be the starting materials for another.

▼ **Basic feedstocks** have been processed from raw materials by, for example, distillation and cracking of crude oil.

▼ **Commodity chemicals** are the next generation in terms of processing and are traded in large tonnages, say from millions of tonnes to several thousand tonnes per year worldwide. They are normally made by a number of different companies and are usually worth less than £500 per tonne. They are the key building blocks for the chemical industry. Examples include ethene, benzene, methanol, sulfuric acid and quicklime (calcium oxide).

▼ **Commodity/specialised** chemicals have been subjected to more processing and are traded in somewhat smaller tonnages, say a few thousand tonnes per year world wide.

▼ **Fine chemicals** are normally of high purity and are traded in tonnages of between 10 and 1000 tonnes per year worldwide. Their values range from £5 to £100 per kg. They have a known chemical identity and are bought and sold on the basis of their chemical specification, *ie* for what they are. Examples include bulk active ingredients for pharmaceutical products and agrochemicals and also flavours and fragrances.

▼ **Speciality chemicals,** sometimes called **performance** or **effect chemicals** are bought and sold for their effect rather than their chemical identity, *ie* for what they do rather than what they are. They are often made by just a few

RS•C

companies in quantities ranging from kilograms to several tonnes per year. They may be formulated (*ie* have a number of ingredients). They include household chemicals (such as bleach or washing up liquid), cosmetics and DIY products like paints. Other examples include colouring agents, thickening agents and stabilisers.

None of these terms is precisely defined and the categories overlap somewhat. Different sectors of the industry tend to use them somewhat differently and other terms such as ultrafine are sometimes used. The basic principle however is that more chemical processing gives substances in smaller quantities but of higher value per tonne.

The picture is summarised in *Fig 3*.

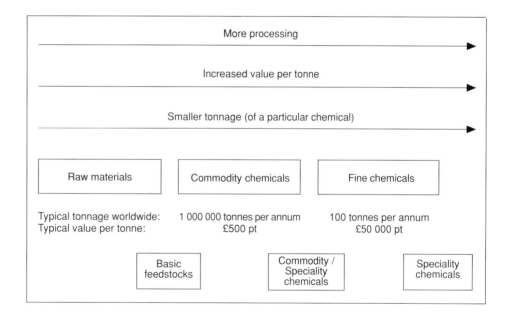

Figure 3 Processing value of chemicals

Selling chemical services – the analytical sector

A further sector of the chemical industry is concerned not with selling chemicals themselves but with selling chemical services, such as chemical analyses. This sector is believed to represent a turnover of about £7 billion per year. Much of this activity is found within organisations whose main business is making and selling chemicals but there is an increasing independent sector which undertakes contract work.

The chemical industry and the environment

In recent years, there has been increasing concern about the environment. There have also been high-profile chemical incidents at Flixborough, UK (1974); Seveso, Italy (1976) and Bhopal, India (1984).

The exposure of people to chemicals is controlled by the Control of Substances Hazardous to Health Regulations (COSHH) (1988) made under the 1974 Health and Safety at Work Act. In 1995 the Environmental Agency (EA) for England and Wales was set up to administer integrated pollution control (IPC). The EA replaced a number

RS•C

of other organisations which had been responsible for different aspects of pollution control. More recently (1996) the government has introduced a landfill tax levied at two rates to encourage better methods of waste disposal and production of less waste. All these pieces of legislation have affected the chemical industry's procedures and methods of waste disposal.

RS•C

RS•C

Bibliography

The UK Chemical Industry, Chemical Industries Association, Published annually. This is a brief digest of statistics which is obtainable from The Chemical Industries Association, Kings Buildings, Smith Square, London SW1P 3JJ.

The Essential Chemical Industry, York: The Chemical Industry Education Centre, 1995. This publication is currently being updated and contains details of a variety of industrial processes along with statistics. It also has an introduction to the chemical industry and a brief historical section.

Britain's Chemical Industry, London: Foreign & Commonwealth Office,1997.

Chemical Industry Data – A brief Guide to the Chemical Industry, London: The Royal Society of Chemistry, 1997.

RS•C

Selling chemical services – analytical chemistry

When we think of the chemical industry, we tend to think of the making and selling of actual chemicals whether it be on the scale of many tonnes or a few milligrams. However, there is another side to the chemical industry, that of providing chemical services. This case study looks at two examples. Intertek Testing Services Environmental Laboratories in St Helens, Merseyside, does analytical work on land which may be contaminated with a variety of pollutants from previous users of the site. Robertson Laboratories (based near Llandudno, North Wales) also specialises in analytical chemistry, including a service, called Wearcheck, which regularly analyses lubricating oil samples from machinery to help to predict breakdowns and to plan maintenance.

Analysis of contaminated land

The UK has a rich and varied industrial past. One unfortunate legacy from this is that many pieces of land are polluted by various chemicals resulting from factories and other industrial sites which are now derelict. This is sometimes called brown land. Some causes of land contamination include:

▼   chemical works;

▼   asbestos works;

▼   gas works;

▼   steel works;

▼   landfill sites;

▼   paper and printing works;

▼   sewage farms;

▼   tanneries;

▼   garages; and many others.

The range of possible pollutants is large and diverse but typically might include:

▼   cyanides from electroplating works;

▼   waste oil and petrol from garages;

▼   coal tar products from gas works;

▼   pesticides and phenols from timber treatment works; and

▼   heavy metals from sewage sludge.

People have not always been as aware of the risks associated with chemicals as we are today nor have appropriate disposal methods always been available, or, where available, followed.

Legislative background

There is no single legal requirement for land to be tested for contamination before it is redeveloped for building, for example. However, there are a number of considerations.

RS•C

▼ Any future problems caused by contamination could result in legal action for damages against the owners of the land or the original polluter. Ill health caused by toxic waste leaking into the voids beneath the floors of houses built on contaminated land could be one example. Another example is explosions caused by methane gas from landfill sites which could destroy houses.

▼ The Environment Agency (which includes what was formerly The National Rivers Authority) may become involved if there is the possibility of watercourses becoming polluted.

▼ Planning Authorities need to be consulted if there is to be building on the site.

▼ The Health and Safety Executive must be informed if there is any potential hazard to workers on the site.

▼ Local authorities are required to identify potentially contaminated land and may require remediation (clean up).

▼ The Department of the Environment, Transport and the Regions (DETR) provides a series of guidelines about investigating contaminated or marginal land.

Not all sites with an industrial history are contaminated, of course. It may be clear that a site is contaminated (because of visible problems such as leaking oil) or because of the site's history (as a landfill site, perhaps). On the other hand a site might be known to be free of problems (one that has always been grazing land, for instance). Many sites are in neither of these categories, though, and an analytical company such as ITS might be approached by a landowner (usually through an environmental consultant) to assess a piece of land for contamination.

The desk study

The first step in an assessment of potentially contaminated land is the desk study. It is so-called because it is carried out away from the site. It involves researching the history of the site using old maps and other records to build up a picture of any industrial activity there. This might involve consulting local libraries and museums, referring to old newspapers or even talking to local residents of long standing such as people who worked on the site some time ago. If it becomes clear that the site has an industrial past, further research is needed to locate records such as engineering drawings which might point, for example, to the location of storage tanks or pipe runs.

The dates when the plant operated are also important, as chemical processes and technology have changed over the years. It may be neccessary to consult old textbooks to find out the details of obsolete chemical processes to get clues as to what chemicals to look for. These include raw materials, products and by-products. By-products, incidentally, are more likely to be found as contaminants, as manufacturers are more likely to take care with the more valuable raw materials and products.

Eventually a site visit will be made to look for clues which can be seen (or even, unavoidably smelled). Occasionally locals walking their dogs have been known to give useful information, sometimes over a pint in the nearest pub!

A desk survey should give an indication of what pollutants to look for – the target contaminants. For example, if the site was formerly a gasworks, likely contaminants include:

▼ metals such as arsenic, cadmium, lead and mercury;

RS•C

▼ sulfur, sulfides and sulfates;

▼ cyanides;

▼ ammonia;

▼ hydroxybenzenes (phenols); and

▼ polyaromatic hydrocarbons (PAHs).

Coal gas   **Box 1**

Until the late 1960s / early 1970s, when it was replaced by methane (natural gas), gas piped to British homes was a mixture of hydrogen and carbon monoxide called coal gas. It was made by heating coal in the absence of air. Coal gas was driven off and coke, ammoniacal liquor and coal tar were left behind. Coal tar is a mixture of hundreds of organic compounds including phenols and polyaromatic hydrocarbons. Thus the last two groups of compounds are often found on old gasworks sites.

RS•C

Polyaromatic hydrocarbons (PAHs)   **Box 2**

Aromatic hydrocarbons have a delocalised system of π-electrons around hexagonal six-membered rings of carbon atoms. The best-known example is benzene. Polyaromatic compounds have two or more of these rings joined by sharing one or more sides, and they retain many of the special properties of aromatic systems. Some PAHs are shown below.

Naphthalene

Anthracene

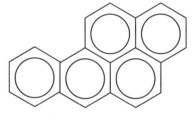

Benzopyrene

Phenanthrene

Examples of polyaromatic hydrocarbons

Polyaromatic hydrocarbons are found in the coal tar produced in old gas works which produced coal gas from coal. A number of them are carcinogens (cancer-causing compounds). A particularly potent one is benzopyrene which is also found in cigarette smoke and car exhaust fumes.

A typical desk study might cost £500 plus and at this stage it might be possible to give an estimate of the probable cost of sampling and analysis to the client. This could vary from £1000 to £100,000 and possibly represent about 1% of the total cost (or 5% of the land remediation cost) of an engineering development.

The sampling process

The desk study and a site visit will have given the analytical chemist some idea of what to look for and possibly where to look for it (based on the locations of old processes). The next stage is to take samples of the soil (and sometimes water and occasionally gas) from the site.

Samples are normally taken on a square grid pattern across the site (unless there is good historical data which enables the most useful sampling points to be targeted).

RS•C

Typically a grid 50 m x 50 m is used although this depends on the budget; a tighter grid generates more samples (and therefore costs more) but is less likely to miss a hot spot of localised contamination. The normal method of sampling is to dig trial pits with a JCB-type digger. The pits are between 3 m and 5 m deep – the length of the digger's arm. Samples are taken every 1/2 m down the pit (or where variations in strata are noticed or there is an obvious change in appearance). This generates a 3-dimensional matrix of samples – *Fig 1*. Samples are taken by an experienced chemist or geologist. She or he observes the sampling and keeps a log (*Fig 2*) plus photographs or sketches of her or his observations. These may provide vital clues such as:

▼   seeing black tar;

▼   seeing a yellow colouration of the soil which might suggest chromium(VI);

▼   detecting characteristic smells;

▼   noting features below ground such as rubbish, suggesting a landfill site; and

▼   observing how quickly the pit fills with water.

Antique objects, bones, fossils and even unexploded bombs have been found during sampling!

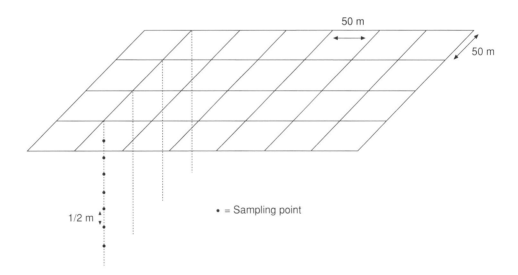

Figure 1 A typical sampling grid – not to scale

RS•C

Intertek Testing Services Environmental	PROJECT REFERENCE	GUIDE		TRIAL PIT	10A
	CLIENT REFERENCE			DATE	- 17/01/97

PLANT JCB 3CX		STABILITY UNSTABLE		GROUNDWATER NONE DETECTED

Samples		Depth m.	Description of Strata	Legend
Type	Depth m.			
		0.20	GRASS COVER, DARK TOPSOIL (0.20)	
S1	0.50		MOIST, GREY SANDY CLAY WITH SANDSTONE FRAGMENTS AND OCCASSIONAL BRICK RUBBLE (1.20)	
S2	1.50	1.40	RED SANDY CLAY WITH SANDSTONE FRAGMENTS (0.40)	
S3	2.00	1.80	YELLOW BROWN SANDY GRAVEL WITH SANDSTONE FRAGMENTS (0.50)	
		2.30		

LOGGED BY MOC
SCALE 1:25
Sample/Test key:
S Spot soil sample
B Bulk soil sample
W Water sample

REMARKS (CHEMICAL ODOURS ETC.) NONE DETECTED

TRIAL PIT PLAN BEARING

Figure 2 Extract from a sampling log

Sampling on site is expensive and it is important that it does not have to be repeated, so more samples are taken than may actually be needed. Samples of about 1–2 kg of soil are taken at each point although only about 70 g are normally required for analysis. The extra is taken to allow for sample preparation to ensure that the sample is the same throughout, *ie* to even out variations. Some may be stored to allow for re-testing. While on site, chemists may request further sampling on the basis of their observations. A pit may be extended into a trench, for example, to measure the extent of observed contamination in a particular direction.

Other sampling techniques include drilling boreholes and examining the cores of earth extracted, and installing piezometers (instruments for measuring pressure) which are used to take readings of water levels and other data. It is important to understand the hydrogeology (water flows) of the site because the flowing water tends to spread pollution in a particular direction (*Fig 3*). The extent of this depends on how water soluble the pollutants are.

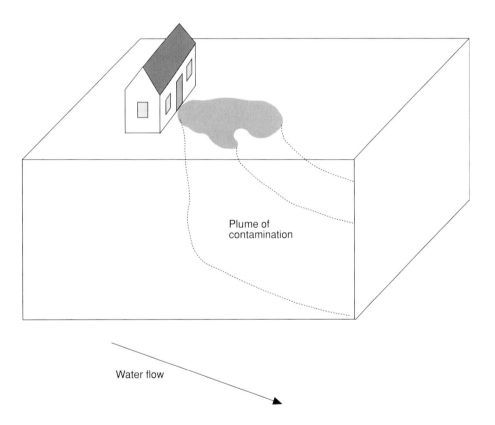

Figure 3 The spread of contamination carried by ground-water flow

Samples must be treated appropriately, and a meticulous labelling and record-keeping system is used so that exactly what has happened to any sample can be traced. This is because the results could possibly be used as evidence in court.

Examples of special treatment of samples include:

▼   samples containing volatile materials must be stored in containers with no headspace or the volatiles could evaporate into the air inside the sample jar;

RS•C

▼   samples to be tested for ammonia (which is volatile) are preserved by adding a small amount of sulfuric acid which fixes the ammonia present; and

▼   samples to be tested for cyanide ions are placed in sodium hydroxide to prevent the formation of hydrogen cyanide gas which could escape.

Question 1

(a)   Write an equation for the reaction of ammonia with sulfuric acid solution. Use it to explain why this fixes the ammonia (prevents loss of ammonia from the sample).

(b)   Write an equation for the formation of hydrogen cyanide gas from the reaction of cyanide ions with water. Explain how the addition of sodium hydroxide prevents this reaction.

The chemical analyses

It is not possible to carry out one single type of analytical test which determines all possible contaminants qualitatively (what is present?) and quantitatively (how much is present?). A variety of different techniques is needed. With good judgement and sound chemical knowledge, the desk survey and site visit gives the chemists a good idea of what contaminants to look for. The aim is to assess the site with as few tests as possible to minimise the cost to the client.

For example, one possible group of contaminants of many sites is hydroxybenzenes (phenols, *ie* compounds with an OH group attached directly to an aromatic ring). One straightforward test which indicates the presence of phenols is to extract the soil sample with methylbenzene (toluene) by Soxhlet extraction (see Box 3) to dissolve the phenols and other organic compounds. The phenols are then concentrated by extracting the methylbenzene solution with an aqueous alkali, and the reagent 4-aminophenazone is added. The phenols dissolve preferentially in the alkaline solution and give a reddish colour when 4-aminophenazone is added. The intensity of this colour can be measured at a specific wavelength with a UV/visible spectrophotometer (a sophisticated colorimeter). The greater the absorption of light, the more phenols were present in the soil sample.

RS•C

Soxhlet extraction   Box 3

This is a method of extracting compounds from a solid sample into an organic solvent. The ground solid sample is placed in a porous thimble in the apparatus (see below).

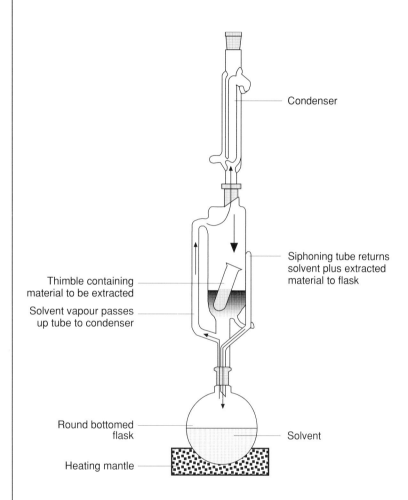

Condenser

Siphoning tube returns solvent plus extracted material to flask

Thimble containing material to be extracted

Solvent vapour passes up tube to condenser

Round bottomed flask

Solvent

Heating mantle

Soxhlet apparatus

The solvent boils and its vapour condenses in the condenser and hot, liquid solvent drips through the sample in the thimble, dissolving any soluble material. Every so often, the solvent siphons back into the boiling flask, carrying with it the extracted material. The process is continued until all the soluble material has been extracted into the solvent. Arrows show the flow of solvent vapour (upwards) and liquid solvent (downwards).

Question 2

Point out some features of the method which ensure that all the soluble material is extracted as quickly as possible.

RS•C

Question 3
Explain why

(a)   Phenols tend to dissove better in an aqueous alkaline solution than in methylbenzene.

(b)   How this technique helps to make a more concentrated solution of the phenols.

(c)   Why a more concentrated solution makes the analysis more sensitive.

Question 4
Explain the differences in chemistry you would expect between a phenol (where the OH group is attached directly to the benzene ring) and a compound where the OH group is attached to an alkyl group which is bonded to the benzene ring. Use the two isomeric compounds 2-methylphenol (1) and phenylmethanol (2) to illustrate your answer. Which dissolves better in aqueous alkali?

(1) 2-Methylphenol   (2) Phenylmethanol

Although it is not equally sensitive to all phenols, the test gives an indication of the total quantity present. The test is quick and cheap (about £15 to the client) and for some purposes, this is all that is needed. If the types and amounts of different phenols are required, these must be separated by High Performance Liquid Chromatography (HPLC) (Box 4). A chromatogram like the one shown in Box 4 is produced with each peak representing a different phenol. The retention time indicates which phenol has been separated and the area of the peak its amount. Peak areas can be measured by the computer which collects and stores the data. This is, of course, a more time-consuming and expensive test.

RS•C

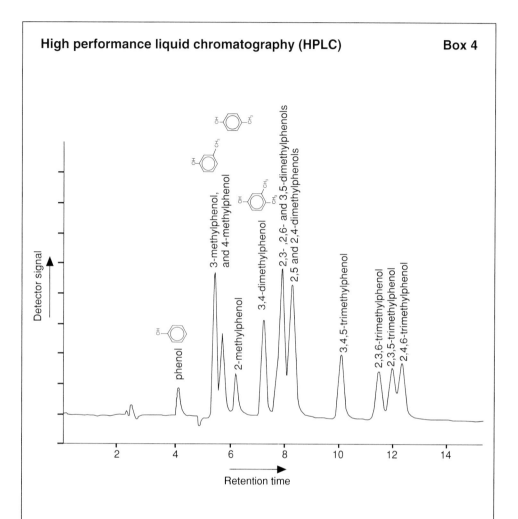

High performance liquid chromatography (HPLC)   Box 4

HPLC chromatogram for separation of phenols

In this technique, mixtures are separated by being forced along a column of solid material by a solvent which is pumped through the column under high pressure. (HPLC is sometimes taken to stand for high *pressure* liquid chromatography.) Components of the mixture travel along the column at different rates which depend on their affinity for the column material (called the stationary phase). The greater the affinity, the slower they travel. The time taken by a component to traverse the column is called the retention time; it is characteristic of a particular compound and can be used to identify it. Some of the structures are shown; try to work out the others from their names.

The Pathfinder package

Many customers prefer to use an off-the-shelf package of tests to screen their samples. ITS has developed its Pathfinder package for this purpose. It includes tests for:

▼   metals – As, Cd, Pb, Hg, Cu, Ni, Zn, Cr, Se, [B] (boron is a semi-metal, but is included here as it is determined along with the metals);

▼   phenols;

▼   the ions CN^-, S^{2-}, SO_4^{2-};

▼   pH; and

RS•C

▼   methylbenzene (toluene) extract which is a simple measure of oil, tar and grease which dissolve in methylbenzene.

Question 5

Explain why oil, tar and grease dissolve better in methylbenzene than in water.

After interpreting the results of these tests, it may be necessary to do others. For example, if a high value of total lead (in all forms) is found on, say, the site of an old petrol station, it may be worth doing a determination of organic lead (Box 5) because the lead-based anti-knock agent in leaded petrol is tetraethyl lead ($Pb(C_2H_5)_4$). Again there is an economic consideration. The total lead value is determined (along with other metals) by inductively coupled plasma emission spectrometry (ICPES) (Box 6) and costs about £5 per test. The organic lead test involves extracting the lead compounds into an organic solvent (in which organic lead compounds but not inorganic lead compounds are soluble). They are then extracted into nitric acid and measured by atomic absorption spectrometry (AAS) (Box 7) at a cost of about £50 per test. If all the individual lead compounds are to be separated and measured individually, this must be done by gas chromatography/mass spectrometry (GCMS) at typically £100 per test.

Organic and inorganic lead   **Box 5**

Lead is in Group IV of the Periodic Table, the same as carbon. It can form compounds in oxidation state +II, which are generally ionic, and oxidation state +IV which are generally covalent. Organolead compounds are those in which lead forms covalent bonds with carbon atoms. These contrast with inorganic lead compounds which have no lead-carbon bonds. For example, in lead(II) carbonate, $PbCO_3$, the bonding is ionic, (Pb^{2+}, CO_3^{2-}) even though the compound contains both lead and carbon so this compound is considered to be inorganic.

RS•C

Inductively coupled plasma emission spectrometry (ICPES)   Box 6

Inductively coupled plasma emission spectrometry is a form of emission spectrometry and is a sophisticated variant of the simple flame test. When atoms are excited, electrons are promoted to higher levels and then fall back to lower levels giving out quanta of electromagnetic radiation as they do so (see below). The frequency of the emitted radiation is linked to the energy gap between the two electronic energy levels concerned. Each element has a unique set of energy levels and therefore emits a characteristic pattern of frequencies which can be used to identify it. By measuring the intensity of the radiation given out it is possible to determine the element's concentration.

In ICPES, the sample is excited by being sprayed into an argon plasma (a highly ionised gas) at a temperature of up to 10 000 K. The plasma is generated by applying a radio frequency field to a flow of argon gas which has been heated to make it conduct. The oscillating magnetic field of the radio signal induces electric currents in the argon which heat it to sufficiently high temperatures to vaporise, atomise and excite the sample. The radiation emitted from the sample is recorded and analysed by computer to identify and measure the levels of elements present.

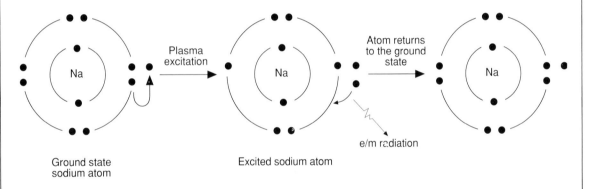

Ground state sodium atom   Excited sodium atom

Emission of light from an atom during ICPES, using sodium as an example

---

Atomic absorption spectrometry (AAS)   Box 7

This analytical technique is often used to measure small concentrations of metals. All elements absorb electromagnetic radiation at specific wavelengths which correspond to the gaps between their electronic energy levels. In AAS, a solution containing the metal of interest is sprayed into a flame. This vaporises the sample and converts it into separate atoms. Light of exactly the right wavelength to be absorbed by the metal in question is shone from a lamp through the flame. The more metal in the flame, the more light is absorbed and thus the concentration of the metal in the original sample can be determined.

Some further examples of analytical techniques and what they can be used for are given in Table 1.

RS•C

Substance(s) to be measured	Technique
Heavy metals	Inductively coupled plasma emission spectrometry (ICPES) or atomic absorption spectrometry (AAS)
Chlorinated organic compounds – *eg* polychlorinated biphenyls (PCBs)	Gas chromatography (GC) with electron capture detection
Aromatic compounds	Gas chromatography-mass spectrometry (GCMS) or GC with flame ionisation detection
Polyaromatic hydrocarbons	High performance liquid chromatography (HPLC) or GCMS
Cyanides	Titration with silver nitrate
Petroleum hydrocarbons	Infrared spectroscopy (IR), GC or thin layer chromatography (TLC)

Table 1 Analytical techniques used to measure various species

Remediation of the land

Once the chemical analysis is complete, ITS reports back to the client. The report is a factual one, indicating what chemicals have been found in what levels and where on the site. Further interpretation and the decision as to what to do next is the responsibility of the client or his consultant. The report confines itself to those substances that the client has requested although, in some cases, the analyst may draw unusual results to the attention of the customer.

There are a number of possibilities for remediation (clean up) of contaminated land. Which is chosen depends on cost, the detail of the problem (landfill gas, watercourse pollution, *etc*) the detail of the pollution and the **end use** of the land. For example, a site destined for use as a car park might be deemed to need less clean up than, say, a domestic garden where vegetables might be grown. In general, in the UK, the policy is to clean up land to a condition which is suitable for the intended use, rather than to a green field condition.

Question 6

(a) Suggest as many reasons as possible why the site for a car park might be thought to need less clean up than the site for a leisure park or house.

(b) Another approach is that *all* sites are required to be cleaned up so that they are pristine, irrespective of the end use. This is called **multifunctionality** in contrast to the end use approach. List the arguments for and against the two approaches.

Some possibilities for clean up include:

▼ Digging up the affected soil for disposal in a landfill site elsewhere and replacing it with unaffected soil.

RS•C

▼   Containing the pollution on site. This can be done by sealing it within a clay, plastic or concrete barrier, for example. This might prevent chemicals leaching out into a watercourse. A higher-tech version of this is an electro-kinetic fence. Here electrodes are sunk into the ground. Ions of pollutants are attracted to the electrodes where they react chemically.

▼   Chemical treatment. For example chromium(VI) salts could be reduced to less-toxic chromium(III) by reduction with iron(II) salts.

▼   Volatile compounds can be removed by vacuum degassing. This is done by sinking boreholes and pumping out air. Volatile compounds vaporise and can be burnt, trapped or oxidised using a catalytic converter.

▼   Bioremediation – this involves using microorganisms to speed up natural degradation of certain chemicals.

▼   Heavy metals can be converted to their sulfides which are insoluble and therefore are not leached into water courses.

▼   Steam stripping involves blowing steam through the affected soil to vaporise and carry away some of the pollutants.

▼   Vitrification involves chemically treating soil and heating it to convert it into an inert, insoluble, glassy substance which traps pollutants.

▼   Soil washing involves concentrating the contaminants by a combination of soil separation (many contaminants are found preferentially on small particles) and cleaning the soil with water and detergents.

Sometimes the best option for dealing with some contaminated land may be to leave it alone or enclose it.

Question 7

Write a balanced equation for the reaction of chromate(VI) ions (CrO_4^{2-}) with iron(II) ions (Fe^{2+}). Assume the products are Cr^{3+} ions and Fe^{3+} ions. What other reactant would be required? What colour change would you see?

Solubility and pollution	Box 8
Many metal ions are toxic to humans. However, they can only show this toxicity if they can get into the bloodstream. To do this they must be water-soluble. A good example of this is the metal barium. Barium chloride (solubility 1.5 mol dm^{-3}) is highly toxic because of the barium ions it contains, but barium sulfate (solubility 9.5×10^{-6} mol dm^{-3}) is so safe that it can be deliberately given to patients as part of a barium meal to help in X-raying the digestive system.	

RS•C

Wearcheck

We often talk of *preventive* maintenance – the regular servicing of machinery so that potential problems can be rectified before they become apparent. For example, a worn fan belt in a car engine can be replaced before it actually breaks. Wearcheck is about *predictive* maintenance – actually forecasting *when* and *how* a piece of machinery is likely to break down so that parts can be replaced at a convenient time *before* failure. This seemingly magical feat can be achieved by sampling the lubricating oil of machinery at regular intervals.

If a particular part of a machine, such as a bearing, is wearing, the metal that wears away is carried away into the lubricating oil which bathes the moving parts as it runs. The concentration of metal debris in the oil can be measured and compared with a baseline level which has been established for that machine. Increased amounts of metal debris suggest abnormal wear. The technique can go further than this; different components of a machine – pistons, seals, bearings *etc* – are made from different metals and alloys so abnormal levels of a particular metal can pinpoint the component where wear is occurring. Some examples are given in Table 2.

Element		Significance
Aluminium	Al	Accessory drives
Aluminium	Al	Dirt
Aluminium	Al	Piston/bearing
Aluminium	Al	Thrust washers
Boron	B	Coolant
Chromium	Cr	Hydraulic rods
Chromium	Cr	Ring, seals and bearings
Copper	Cu	Bearing/bushings
Copper	Cu	Clutch discs
Copper	Cu	Thrust washers
Copper	Cu	Turbo & cooler
Lead	Pb	Bearing
Lead	Pb	Cooler & petrol
Lithium	Li	Grease
Manganese	Mn	Castings

Element		Significance
Manganese	Mn	Steel components
Molybdenum	Mo	Ring and seals
Nickel	Ni	Compressor tube
Nickel	Ni	Steel component
Phosphorus	P	Phosphorbronze
Potassium	K	Coolant
Silicon	Si	Clutch disc
Silicon	Si	Dirt
Silver	Ag	Bearings
Sodium	Na	Coolant
Tin	Sn	Bearings
Titanium	Ti	Dirt & paint
Vanadium	V	Fuel oil
Vanadium	V	Steel components
Vanadium	V	Valve stems

Table 2 Contamination and wear elements

This means that a service engineer can often be told exactly which component is to be replaced without having to strip down and examine the whole piece of

RS•C

machinery. It is often possible to predict from the *levels* of different metals in the oil *when* the machinery is likely to fail – sometimes to within a few hours. *Figure 5* shows the change in the levels of various metals caused by dirt in the cylinder of an engine causing excess wear, and the return to normal levels after repair.

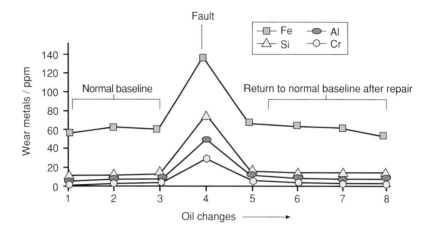

Figure 5 Wear metals before, during and after repair of a fault

The financial saving of this type of monitoring can be considerable. For example, an out-of-service underground mining machine can mean a loss of production of up to £50,000 per day and the workforce having to be sent home.

Measuring wear metals

The main technique used for measuring the concentrations of wear metals is inductively coupled plasma emission spectroscopy (ICPES) (Box 6).

In fact ICPES does not measure *all* of any metal present in the oil – it determines only particles less than 3 μm in size (because larger ones do not vaporise in the instrument's plasma). Oils may contain particles of up to 30 μm. However failure prediction is based on *changes* in metal concentrations rather than absolute values. Increased wear produces an increase in metal particles of all sizes and so an increase in the levels measured by ICPES reflects an overall increase in wear.

Further analyses of oil

Further analyses of oil can indicate other problems as well as wear including:

▼   dirt in the oil;

▼   fuel in the oil;

▼   water in the oil;

▼   antifreeze in the oil;

▼   acidity or basicity of the oil;

▼   ferrous (iron/steel-based) debris in the oil; and

▼   deterioration in the oil itself.

RS•C

Dirt

Dirt in engines leads to excessive wear. Detecting dirt may lead to a *prediction* of excessive wear later or it may *explain* excessive wear found by metal analysis. Dirt in the oil from road or concrete dust, for example, usually contains silicon from silicates. Silicon levels are measured, along with wear metals, by ICPES.

Fuel

Fuel (petrol or diesel) may be present in oil because of leaks or faulty fuel injectors. It reduces the viscosity (treacliness) of the oil making it less effective as a lubricant and hence causing wear. Fuel in the oil is indicated by the results of two measurements – the viscosity of the oil and its flashpoint (the lowest temperature at which a liquid will give off sufficient vapour from its surface to ignite in air when a spark or flame is applied). Fuel in the oil lowers its flashpoint. Viscosity is measured by timing how long a sample takes to flow under gravity through a capillary tube. Flashpoint is measured by the open cup method in which a sample is gradually heated and every two degrees of temperature rise a small flame like a gas pilot light is applied to the surface to see if the oil ignites. A low viscosity alone does not nesessarily indicate fuel contamination – it could be caused by deterioration of the oil itself

Water in the oil

The presence of water can be shown very simply by the crackle test. A little oil is placed on a hotplate at over 100 °C; if water is present it will crackle as the water boils – think about putting wet chips into hot cooking oil. If water is present its concentration can be measured by distilling it from the oil.

Water can be caused by leaks of coolant into the engine or by contamination in the oil storage tank (a tank open to the rain, for example). If a significant amount is found (>0.2%), its source must be determined. If a coolant leak is the cause, traces of the coolant antifreeze (usually ethylene glycol, ethane-1,2-diol) and its additives will also be present. This can be tested for by oxidising the glycol to ethanal by periodate followed by reaction of the ethanal with Schiff's reagent which produces a magenta colour suitable for colorimetric determination. However, some oil additives interfere with this test and it is often better to try to measure the anticorrosion additives in the antifreeze. These contain the elements sodium, potassium, silicon and boron which can be measured by ICPES.

Even so, results must be interpreted with care. Small amounts of water contamination can evaporate away at the temperature of the engine oil, and glycol can be oxidised in the engine to oxalic acid (ethanedioic acid). If this is the case, no water or coolant may be detected even though a leak is actually present.

Acidity and basicity of oils

Most fossil-based fuels contain small amounts of sulfur. On combustion, this forms the acidic oxides sulfur dioxide and sulfur trioxide. The combustion process also results in the combination of nitrogen and oxygen from the air to form a mixture of acidic nitrogen oxides, usually called NO_x. These are the same processes which contribute to acid rain. Unless removed, these acids corrode the inside of an engine. Therefore lubricating oils have additives which are able to neutralise these acids. The amount of acid that these additives can mop up can be found by titrating a sample of oil with either hydrochloric or perchloric acid using an automated titration technique.

RS•C

Question 8

What problem is likely to be encountered when titrating alkali dissolved in oil with an aqueous acid such as hydrochloric acid solution? Suggest a method of overcoming this problem.

Acids may also be formed in oils by oxidation of the oil itself. These acids can cause the oil to polymerise and therefore thicken, affecting its lubricating properties.

Question 9

Lubricating oils are long chain hydrocarbons. What type of acids might be formed by their oxidation?

Ferrous debris

This is an indication of general engine wear, as the majority of a typical engine is made of iron and steel (ferrous metals). The total amount is measured by measuring the effect of these magnetic particles on an applied magnetic field.

A number of other tests are carried out on the oil including a general visual check on appearance to look for large particles of dirt or other debris and for the presence of water in the oil. Examination under a visual microscope can indicate the size and shape of any particle. The source of these may then be able to be determined by a trained diagnostician.

Sampling

Without careful sampling, the Wearcheck system would be useless. If the oil is sampled and sent for analysis in a dirty, wet or rusty container, this will obviously affect the levels of contaminants measured. Robertson therefore supplies customers with its own sampling kits as part of the service. To encourage customers to use them, samples received in non-standard containers are charged extra.

Interpreting the results

The raw results of these chemical tests, such as the oil contains 3 ppm (parts per million) of aluminium or it contains 3.1% of water would probably be meaningless to most operators of machinery. What they want to know is whether there is any cause for concern and, if so, what they should do about it. Robertson employs diagnosticians to interpret for their customers what the results mean. This involves understanding the possible sources of various elements and their possible significance. Often the measured levels are compared with a normal baseline which has been established for that equipment and a particular lubricating oil. Diagnosis is a skilled job and requires a knowledge of both chemistry and engineering. Once the source of any abnormal results has been determined, it may be possible to suggest when a failure is likely and what maintenance work is needed to prevent it. Reports to the customer are then sent by e-mail or fax (if urgent) or by post. Two typical record forms (*Figs 6* and *7*) are shown. The first shows normal readings, the second indicates a potential problem. Notice how specific the recommendations for remedial work are.

RS•C

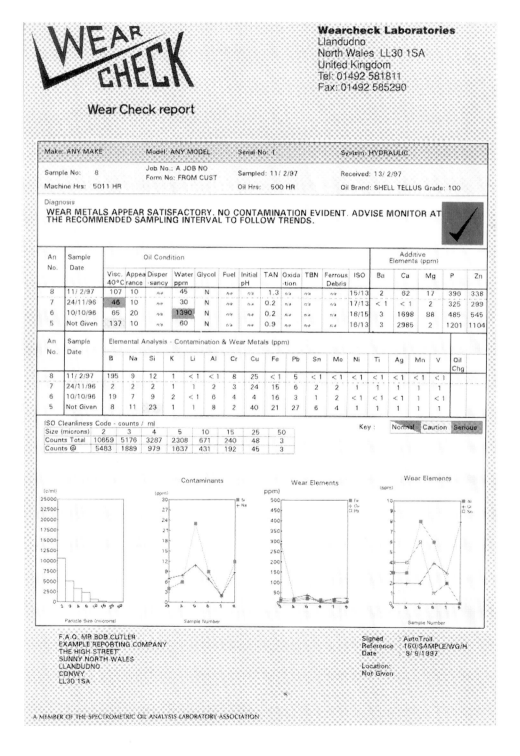

Figure 6 A typical Wearcheck report showing normal readings

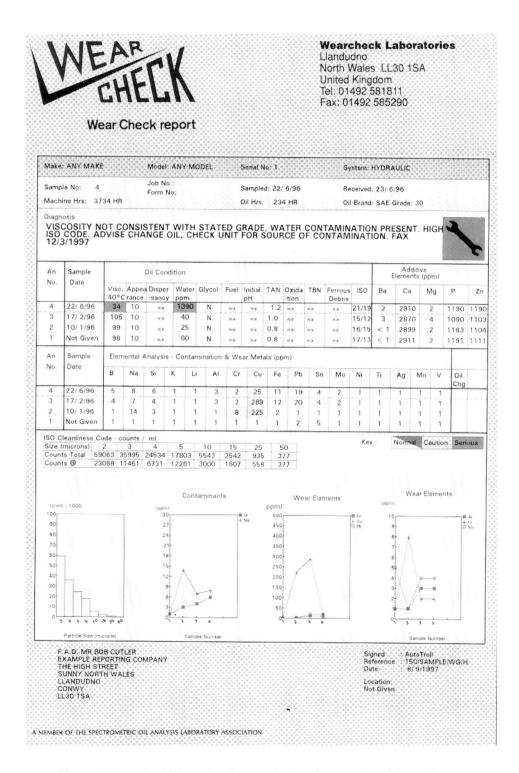

Figure 7 A typical Wearcheck report showing a potential problem

RS•C

Record forms

Some customers prefer to have routine results sent to them on floppy disk to be downloaded into their own computers or even sent to them on-line by modem from the Robertson computer. This enables them to plot graphs showing long term trends for individual pieces of machinery or whole fleets

---

Quality control   **Box 9**

Sample spiking

Quality control is vital in an analytical laboratory where the accuracy of results is all important. One way of checking the analytical process is by spiking the samples which are tested. One type of spiking involves adding to the test sample a known amount of a similar but distinguishable compound to the one being tested for. For example, in a test for methylbenzene (toluene) a spike of deuterated methylbenzene (in which all the hydrogen atoms are replaced by deuterium atoms, ^2H) could be added. Chemically this behaves identically to normal methylbenzene but can be distinguished by its greater relative molecular mass. If all is well, the analytical test should detect the known added amount of deuterated methylbenzene.

Question 10

If all the hydrogen atoms in a molecule of methylbenzene have been replaced by deuterium, how much greater will be the relative molecular mass? Suggest a spectroscopic technique that can measure this mass difference.

Testing the testers   **Box 10**

It is important that the results of all laboratories doing analytical work are comparable and accurate. To ensure this, some organisations make up reference materials – samples to which known quantities of materials to be analysed for (called analytes) have been added. These are sent to various laboratories and their results are compared with the known amount of analyte added to the sample. Only laboratories whose results are within an acceptable margin of the known value are allowed to tender for analytical work for this organisation.

RS•C

Answers to questions

1. (a) $2NH_3(g) + H_2SO_4(aq) \rightarrow (NH_4)_2SO_4(aq)$

The volatile ammonia reacts with sulfuric acid to form ammonium sulfate a non-volatile ionic compound which remains dissolved in the water.

(b) $CN^-(aq) + H_2O(l) \rightleftharpoons HCN(g) + OH^-(aq)$

Addition of hydroxide ions pushes the equilibrium over to the left, suppressing the formation of HCN gas.

2.   The solvent is hot (almost at its boiling temperature). A continuous supply of fresh solvent drips through the solid sample. The sample is ground up, giving it a large surface area.

3. (a)   Phenols are acidic and react with sodium hydroxide ions to give the salts such as sodium phenoxide (Na^+ PhO^-). Being ionic, this dissolves in water rather than in methylbenzene.

(b)   If the methylbenzene is extracted with a smaller volume of aqueous alkali, this effectively increases the concentration of the phenol.

(c)   A more concentrated solution absorbs more light and this absorbance is easier to measure.

4.   2-Methylphenol is a phenol as the OH group is attached directly to the benzene ring. This makes the compound acidic because the charge in the 2-methylphenoxide ion (formed by loss of H^+ from the -OH group) can be delocalisd onto the benzene ring making this ion more stable. No such effect occurs in phenylmethanol as the OH group is not attached to the benzene ring. The compound behaves like an alcohol and is only very weakly acidic. 2-Methylphenol will therefore dissolve better in aqueous alkali.

5.   This is an example of the like dissoves like rule of thumb. Non-polar compounds dissolve in non-polar solvents which have similar intermolecular forces (van der Waals forces).

6. (a)   A variety of answers is possible. Some suggestions might include:

▼   people spend less time in a car park than in a house or leisure park;

▼   crops are not grown on a car park;

▼   a car park is normally concreted over; and

▼   children will probably spend less time in a car park than in a house or leisure park.

(b)   Again a variety of arguments is possible, including:

▼   cost – a clean up to pristine conditions is more expensive;

▼   a requirement for pristine clean up might lead to the land not being used and therefore not cleaned up at all; and

▼   partly cleaned up land might need further work in the future if its use changes.

RS•C

7. $CrO_4^{2-} + 3Fe^{2+} + 8H^+ \rightarrow Cr^{3+} + 3Fe^{3+} + 4H_2O$

H^+ ions (acid) are required.

A colour change of yellow Cr(VI) to green Cr(III) would be seen. Fe(II) is green and Fe(III) is brown, but these colours are less intense.

8.   Time will be required for the alkali in the non-aqueous (oil) layer to dissolve in the aqueous layer where the reaction with the acid will occur. This could make the titration slow and the end point difficult to detect. Possible solutions might be:

▼   to shake the oil with water before titration to extract all the alkali into the water (assuming it is water soluble); and

▼   to react the oil with excess acid and then back titrate with standard alkali to determine how much has been used.

9.   Long chain carboxylic acids are the most likely suggestion.

10.   Eight units heavier. Mass spectrometry could measure this difference.

RS•C

Formulating pharmaceutical products

You may know, or know of, someone who is a diabetic (Box 1). Many diabetics can control their condition by diet, others have to inject themselves regularly with insulin. But have you ever thought why they must go to the trouble and discomfort of *injecting* insulin rather than taking tablets or capsules?

The answer is straightforward. Insulin is a protein and, if taken by mouth, it would pass into the stomach where it would be exposed to a cocktail of chemicals including many proteolytic enzymes (which break down proteins) and also hydrochloric acid. These chemicals would break the peptide bonds holding together the amino acids which make up insulin, and the insulin molecule would soon become a mixture of amino acids. It must therefore be injected directly into the body, by-passing the inhospitable chemical conditions in the stomach.

Diabetes	**Box 1**

Diabetes affects about 2% of the population of the UK. It is caused by lack of insulin – a hormone which controls the level of glucose in the blood. Short term symptoms can include coma, thirst, hunger, excessive urination and blurred vision. Longer term effects may be kidney problems, heart disease and blindness. The condition can be controlled and, if treated, sufferers can lead a normal life.

Medicines and drugs	**Box 2**

There is often confusion about these two terms, and the word drug is now often associated with illegal substances such as ecstasy, cocaine and heroin.

Strictly speaking, a drug is a substance which affects how the body works – either for better or worse. A medicine is a substance that improves health. Medicines contain beneficial drugs (which are the active ingredients) as well as other substances which make them easy and convenient to take.

A newly synthesised chemical which may be a drug and may eventually form the active ingredient of a medicine is often called a New Chemical Entity (NCE).

This is an example of the sort of issue which confronts formulation chemists in the pharmaceutical industry, who may be concerned with the problem of getting an active drug to its target organ in the body in a suitable dose without it being broken down.

Question 1

a)   Draw the structure of a part of a protein molecule showing one of the amide (peptide) linkages.

b)   What is the equation for the breakdown in acid conditions of the amide linkage?

It is almost never the case that a drug can be taken by a patient as the neat active ingredient. Formulation is the process of converting the drug into a form which can be taken by the patient.

RS•C

Ways of administering drugs   Box 3

The following methods are all used to administer drugs.

▼   Oral – tablets, capsules, liquids (including suspensions and solutions)

▼   Injections (intravenous, intramuscular, subcutaneous, intradermal – see below)

▼   Suppositories

▼   Inhalers

▼   Sublingual (dissolving under the tongue)

▼   Skin patches

▼   Nasal drops/sprays

▼   Topical – to the skin via creams, ointments and gels

▼   Pessaries

▼   Dusting powders

▼   Eye drops and inserts

▼   Implants

▼   Paints.

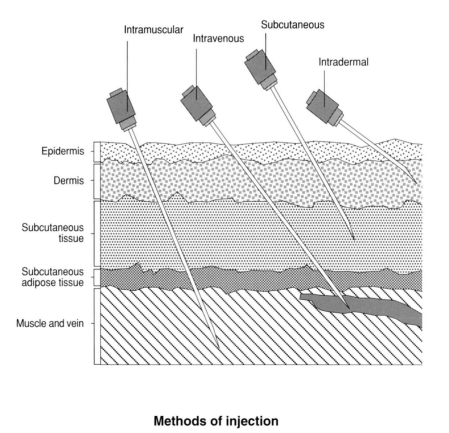

Methods of injection

RS•C

Question 2

Select one of the methods in Box 3; try to think of a medicine which is administered in that way and discuss the pros and cons of this method.

Considerations involved in formulation include:

▼   getting the correct dose into the patient;

▼   releasing the active ingredient;

▼   getting the active ingredient to the target organ;

▼   getting and sustaining the correct level of the drug in the patient's bloodstream, where necessary;

▼   convenience and acceptability to the patient;

▼   shelf life of the drug, *ie* the stability of the drug during storage;

▼   methods of analysis suitable for quality control on the production line; and

▼   ease of manufacture.

Before formulation can start, a thorough knowledge of the physical and chemical properties of the drug is required. Important drug parameters will include:

▼   how soluble is it and how fast does it dissolve at different pHs and in different solvents – especially in the conditions in the gut?;

▼   the stability of the drug at different temperatures, different pHs, with respect to water and in light and whether the drug is hygroscopic (*ie* does it absorb water from the atmosphere?) – these will all affect the best form in which to supply the dose and the packaging;

▼   by what route(s) does it decompose (in the body and in its packaging)?; and

▼   what is the best way to analyse for the drug (for use in quality control)?

Chemical modifications to the original drug may be considered. For example, a drug which is an acid or base might be converted to a salt to make it more or less soluble (for example, the case of erythromycin p. B11). Such modified drugs are called prodrugs – ones which are converted into the active form in the body. Another possibility is that the drug might need to be covered with an acid-resistant coating to protect it from the chemical conditions of the stomach.

Formulation for clinical trials

Formulation chemists become involved with a new drug before the stage of clinical trials, as an appropriate formulation has to be devised for the trials. So a team of formulation chemists will be at work on the drug while the chemical development team is devising the best method for making it. During early trials, the optimum dose will not yet be known so a method of formulation which allows for flexibility of dose is required.

RS•C

Licensing Box 4

All medicines have to go through a licensing process. A licence has to be granted by the authorities in a particular country before any medicine can be marketed there. The pharmaceutical company applies for a licence by providing data about clinical and toxicological trials, safety, the chemistry of the medicine and how it is to be formulated and marketed.

When a medicine is registered, it is the whole formulation, not just the active ingredient which is registered. This is because the matrix of additives with which the drug is supplied may affect its rate of release into the body and the pathway by which the drug eventually decomposes.

If you look on the package of any drug (including over the counter drugs), you will see the product licence code, a number prefixed by the letters PL.

Another consideration is that many of the trials will be double blind trials (Box 5) where neither the patient nor the doctor knows whether an active medicine or a placebo (dummy dose) has been given. It is important that the formulation of the medicine gives no clues as to which is which. So doses of placebo and of active medicine must look and taste the same, be the same colour and have the same mass.

Medicines often have a bitter taste which is particularly noticeable in solutions or syrups to be taken by mouth. Matching the taste of a placebo solution to the active one can therefore be quite difficult. Denatonium benzoate (Bitrex) or sucrose octa-acetate, both bitter-tasting chemicals, can be added to the placebo to mask any bitter taste of the medicine. (Incidentally, these chemicals are added to methylated spirits to deter people from drinking it.) Capsules (gelatin shells into which powdered ingredients can be filled) are one way of blinding the ingredients for this type of trial. Capsules also allow comparison with competitors' products, enabling the rival product to be ground up and put into a capsule for a double blind trial.

Double blind trials Box 5

These are common when testing new medicines. To ensure meaningful results, a set of patients with a particular condition is divided at random into two groups, one of which receives the medicine on test and another which receives the standard treatment or no treatment at all. The no treatment group is given a dummy medicine, called a placebo, which has no active ingredient. Neither the doctor nor the patient knows which group a particular patient is in. The test medicine and placebo are packaged identically and are labelled in such a way that only the researchers know which is which. This is designed to eliminate various types of bias such as patients trying (consciously or subconsciously) to please the doctor by doing better on the new medicine. Also, doctors who know they are prescribing a new medicine might treat patients differently or see improvement where there is none because they hope or believe the new treatment to be better than the old.

RS•C

Bioavailability and the therapeutic window

Bioavailability is a measure of the percentage of the drug in the original formulation which reaches the target organ intact. A drug which is taken by mouth may fail to pass through the gut wall at all, be decomposed by acid or enzymes in the stomach or be decomposed in the liver before reaching the blood stream. Generally it is important that the level in the bloodstream is within certain parameters. Too high and the drug may start to cause unacceptable side-effects, too low and it may have no therapeutic effect. The gap between these levels (which varies considerably between different drugs) is often called the therapeutic window (*Fig 1*).

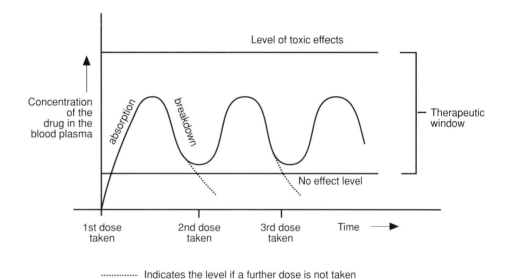

Figure 1 The therapeutic window

RS•C

The case of digoxin **Box 6**

Digoxin is one of a family of glycoside drugs used to treat heart failure. It has a relatively narrow therapeutic window – *ie* the margin between the minimum effective dose and the dose where toxic side effects begin is small.

Some years ago, a number of puzzling deaths occurred due to overdose of digoxin. Subsequent investigation showed that digoxin tablets made by different manufacturers did not have the same bioavailability (they were not bioequivalent), despite containing the same mass of digoxin. Different manufacturers used different fillers in manufacturing their tablets and these were found to dissolve at different rates and led to different peak values of the digoxin concentration in the bloodstream. Patients whose treatment had been stabilised on one manufacturer's tablets could get an overdose if they started to take tablets which dissolved more quickly made by another manufacturer.

Now, the rate of dissolution of tablets is specified as well as the content of active ingredient. Digoxin tablets must be formulated such that 75% of the stated dose dissolves in one hour.

Question 3

What other factors (as well as different fillers) could affect the rate of dissolution of an active ingredient in a tablet?

The structure of digoxin

The actual concentration of the drug in the bloodstream varies with time as shown in *Fig 1*, increasing as each dose of the medicine is absorbed and decreasing as it is decomposed or excreted. Formulation chemists can influence and control this profile by appropriate design of the dosage form. For example, if a drug is absorbed and excreted from the bloodstream rapidly, the patient has to take several doses a day to maintain the concentration within the therapeutic window. This can be inconvenient for the patient, so the formulation chemists can slow down the release of the drug from the dosage form. This is known as sustained release. One way of formulating sustained release medicines is to make them in the form of beadlets consisting of the drug mixed with starch and lactose. Different beadlets are coated with different

RS•C

materials which dissolve at different rates so that some release the drug quickly and others more slowly thus evening out the peaks and troughs in the graph.

Most drugs taken by mouth are absorbed in the small intestine where the pH is between 5 and 6 (*Fig 2*). Many drugs, however, are decomposed in the acidic conditions of the stomach before reaching the small intestine. One solution is to coat the tablet with a so-called enteric coating consisting of a polymer such as ethyl cellulose or a polymethacrylate which is insoluble at pH 1–2 and dissolves at pH 5–6. (Enteric means to do with the intestines.)

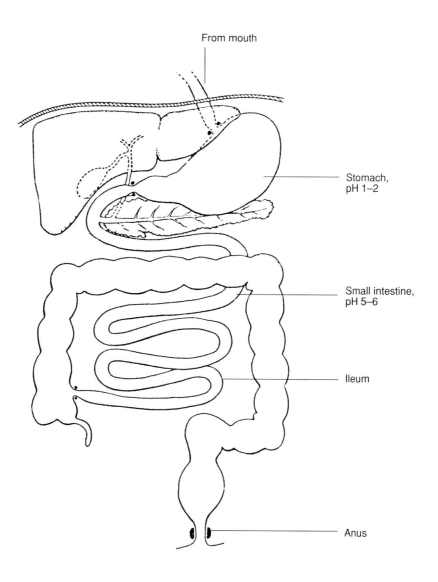

Figure 2 The human alimentary canal

Formulating a tablet

Tablets are the most common way of taking a drug; 50% of medicines are administered by this route. Oral (by mouth) methods are preferred by most patients

RS•C

for medicines which are to be taken systemically (*ie* which enter the bloodstream). Most adults find solid forms of medicines more convenient to take than liquids and such forms also minimise the exposure of the medicine to water, which might hydrolyse it. Tablets are preferred to capsules by patients and are faster, and therefore cheaper, to manufacture.

First, the likely dose must be considered. The practical maximum size for a tablet to be easily swallowed is about 250–500 mg of active ingredient. Doses close to this cause problems, as tablets require a number of additives for manufacturing reasons. Conversely a dose of a few micrograms (typical of many modern drugs) is too small to be handled easily by a patient, so the drug has to be dispersed in an inert **filler** to produce a sensible sized tablet. Depending on its shape a total tablet mass of around 1.2 g is acceptable. Two common fillers (often called **bulking agents** or **diluents**) are the sugars lactose and mannitol (*Fig 3*). Lactose is the more commonly used but mannitol has the interesting property of having an endothermic enthalpy of solution. This makes a mannitol-based tablet feel pleasantly cold if it dissolves on the tongue, and it is often used for chewable tablets.

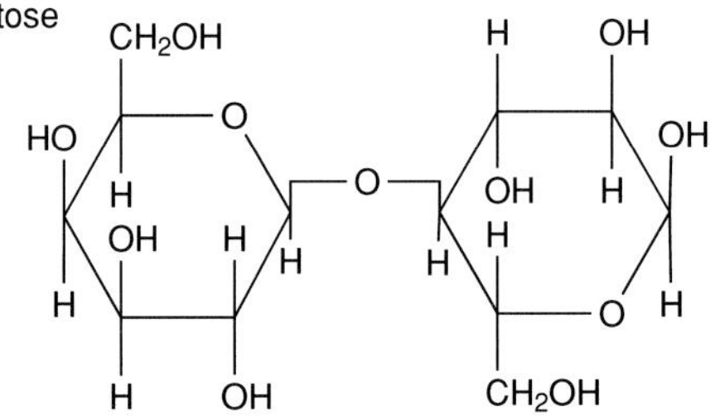

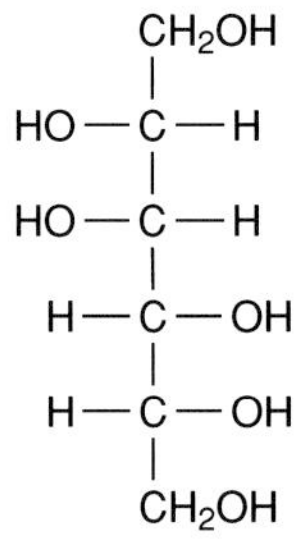

Figure 3 Lactose and mannitol

The active ingredient must be such that it can be evenly dispersed in the diluent to give the same dose of active ingredient in each tablet. The mixture must also compress well into a tablet which sticks together and does not come apart during the manufacturing process or while the tablets are being rattled in the bottle. This can be achieved by adding a **binder** – a cellulose-based solid which is tacky, and holds the tablet together. However, the tablet must break up once in the stomach and it may be necessary to add a **disintegrant** to help the tablet to do this. Typically a starch derivative might be used which swells on contact with water.

During manufacture, the tablet must not stick to the tablet-making machine so a **lubricant** might be added to reduce friction and prevent newly formed tablets sticking to the tablet press. Waxy additives such as magnesium stearate or stearic acid are used as lubricants at a level of about 1%. Too much of these additives could give the tablet a hydrophobic (water repellent) coating which might slow its dispersal in the stomach.

RS•C

Question 4

a)   Starch is a polysaccharide. Give the formula of part of a starch molecule and use this to suggest why starch swells when exposed to water.

b)   The formula of stearic acid is $CH_3(CH_2)_{16}CO_2H$. Why might stearic acid give the tablet a hydrophobic coating?

Hint: both parts of the question concern intermolecular forces.

Packaging

Packaging is an important issue for drug products and may be thought of as an offshoot of formulation. As well as contributing to product appeal and convenience, it has a considerable bearing on shelf life (Box 7). Some examples of packaging issues are:

▼   medicines which are unstable in light need opaque packaging;

▼   medicines which are unstable to moisture can be sold in glass bottles or aluminium foil packs because PVC blister packs are quite permeable to water vapour;

▼   opaque packs. These are thought less likely to encourage children to think that the contents are sweets; and

▼   childproof containers such as push and twist caps.

The decomposition of aspirin   **Box 7**

The problem of shelf life can be illustrated by the case of one of the commonest medicines – aspirin. Aspirin is chemically fairly simple; it is 2-ethanoyloxybenzenecarboxylic acid (2-ethanoyloxybenzoic acid), an ester of ethanoic acid and a phenol. If the tablets are stored in moist conditions for some time, in an unstoppered bottle in a bathroom cabinet for example, then the ester linkage can be hydrolysed leaving 2-hydroxybenzenecarboxylic acid (2-hydroxybenzoic acid) and ethanoic acid. You might have opened an old bottle of aspirins and noticed the smell of vinegar caused by ethanoic acid.

Aspirin hydrolysis

If you come across aspirins in this state, it is best to take them to your local pharmacist (who will dispose of them safely) and buy a new pack.

RS•C

Before registration, tests are done on the medicine in its package at higher than normal levels of:

▼   humidity;

▼   temperature;

▼   illumination; and

▼   oxygen.

These help to predict the behaviour of the medicine and packaging when exposed to normal conditions for a longer time, and are called accelerated tests.

The case of medicine X

Formulation work is an ongoing process and does not stop with the successful marketing of a medicine. It may need to be reformulated for use in a different market, for example in a lower dose for over the counter (OTC) sale rather than prescription-only use. Medicine X is a typical example of these so-called line extensions.

X is a registered medicine which is administered by injection-only to patients in hospital. Injectable medicine solutions must be sterile. This is because the solution enters the bloodstream directly, bypassing the body's defence systems – the skin's barrier and the inhospitable chemistry of the gut, for example. At present X is sold in single use vials (ie each vial contains the amount required for one injection) and the empty vial is then thrown away. It would be cheaper to sell the medicine in a multi-dose vial which contains enough medicine for several injections. This type of vial has a rubber septum cap through which the doctor inserts a syringe and withdraws the appropriate amount for one injection. This means that after the first withdrawal, the remaining solution can no longer be guaranteed to be sterile, as the needle may introduce bacteria. A preservative must therefore be added to kill these microorganisms.

The formulation team for X first conducted a literature search to identify a range of possible preservatives, which had to be safe, odourless and effective against a wide range of organisms at the pH of the injectable solution. Three were selected; phenylmethanol (benzyl alcohol), 3-methylphenol (m-cresol) and 4-hydroxybenzoic acid (4-hydroxybenzenecarboxylic acid, p-hydroxybenzoic acid) (Fig 4).

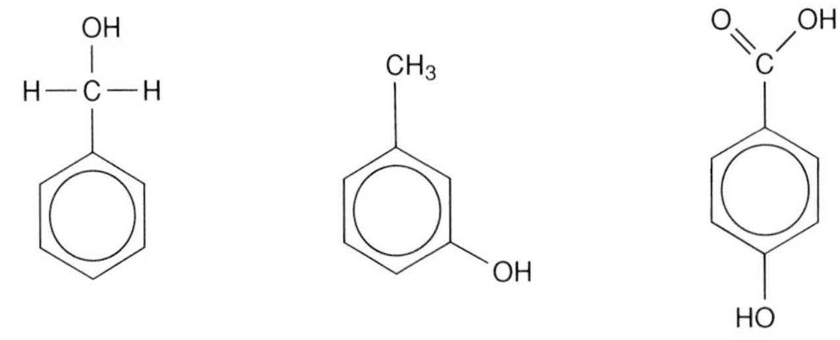

| Phenylmethanol (Benzyl alcohol) | 3-Methylphenol (m-Cresol) | 4-Hydroxybenzenecarboxylic acid (p-Hydroxybenzoic acid) |

Figure 4 Preservatives for drug X

Question 5

Which of the preservatives is *not* a phenol? Explain your answer.

RS•C

The formulators made trial formulations with each of the preservatives. They found that 4-hydroxybenzoic acid was relatively insoluble and needed heat to dissolve it completely; complete dissolution is vital for an injectable medicine. This insolubility was likely to cause manufacturing problems and made it difficult to make up analytical standards. 4-Hydroxybenzoic acid was also found to degrade in the conditions at which it would need to be autoclaved to ensure sterility (121 °C for 30 minutes). This effectively ruled out 4-hydroxybenzoic acid.

The solution of X with added 3-methylphenol was found to contain degradation products of X after autoclaving. Clearly the preservative and X were reacting under these conditions. This ruled out 3-methylphenol.

3-Methylphenol is less used in pharmaceutical formulations than the other two compounds. It also has a strong smell. These factors also mitigated against its use.

This left phenylmethanol, but further tests were still needed. To check that it was effective, a sample of X with added phenylmethanol was "challenged" with a variety of microorganisms to ensure that they were killed within a set time. It was also verified that the formulation of X with added phenylmethanol was:

▼ isotonic with blood plasma (*ie* it had the same osmotic pressure so that there was no tendency for blood cells to shrink or swell by osmosis); and

▼ compatible with the glass and the rubber septum of the proposed vials.

The latter test included accelerated testing with the rubber in contact with the solution at high temperature.

Finally the preservative was tested by making a trial batch of the formulation in the factory (in Puerto Rico) where the solutions were to be made. Here it was found that there was less phenylmethanol in the product than expected; some was being absorbed into the tubing used in the manufacturing plant. This problem was eventually solved by using a different material for the tubing and this formulation of the medicine, in multi-use vials, is on the point of being registered at the time of writing.

Even a relatively small modification to a formulation, such as adding a preservative, requires extensive testing and re-registration of the medicine. The case of digoxin is a classic (and tragic) example, where rates of dissolution caused the problem (Box 6).

The case of erythromycin

Some of the issues that confront the formulation chemist are illustrated by the case of erythromycin (*Fig 5*). This is an antibiotic which has a similar antibacterial effect to that of penicillin but can be used on patients who are allergic to penicillin. It is used against respiratory infections (including pneumonia), whooping cough and legionnaire's disease amongst others.

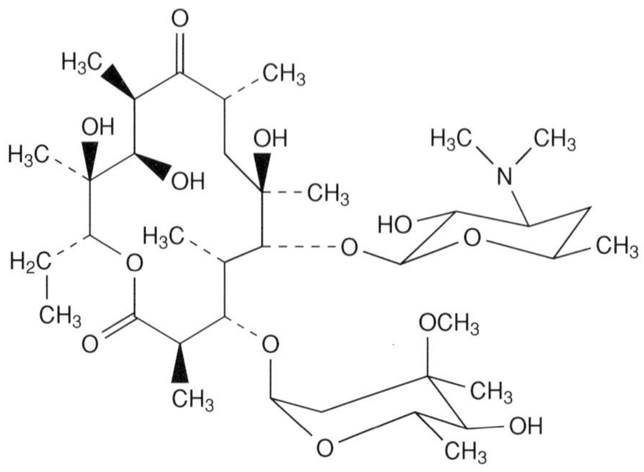

RS•C

Erythromycin. For simplicity this will be drawn as:

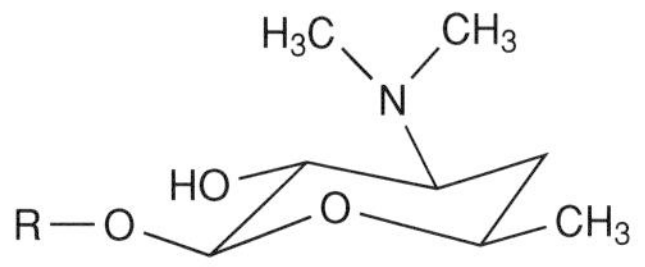

where R =

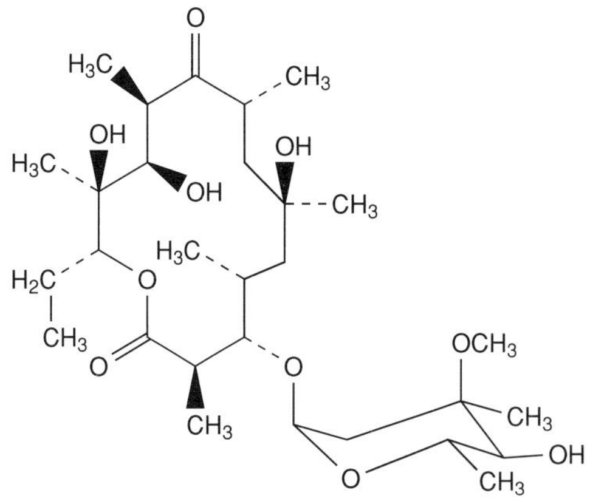

Figure 5 Erythromycin

Erythromycin contains a tertiary amine functional group and is therefore basic. The molecule rapidly breaks down in acidic conditions but is not very soluble at pH 7 or above. It also tastes very unpleasant, having a bitter, metallic taste. These properties present some challenge. We will look at how they are overcome in three situations.

1. The medicine being taken by adults.

2. The medicine being taken by children of under 10 years of age.

3. The medicine being taken by very sick patients who cannot take it by mouth.

Question 6

a) Can you mark the tertiary amine functional group on a copy of the structure of erythromycin?

b) Which bond(s) in erythromycin might break in acidic conditions?

c) Can you identify the chiral centres on the erythromycin molecule? Mark each one with a *. Hint: there are 18.

Overcoming the challenges

1. Adults

Erythromycin is normally taken by adults in tablet or capsule form. This is convenient for storage and for getting the correct dose; no measuring out is needed by the patient. Absorption of the medicine takes place in the small intestine where the pH is between 5 and 6. The small intestine's surface is covered with finger-like villi (*Fig 6*). These are well-supplied with blood vessels and have an enormous surface area making them ideal for absorbing drug molecules into the blood plasma. However, to get to the small intestine, erythromycin must pass through the stomach (*Fig 2*) where the pH is 1–2 (conditions which would decompose it).

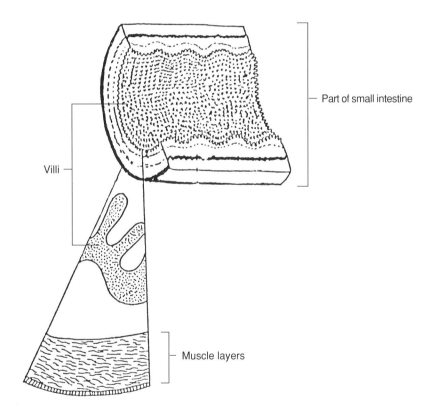

Figure 6 Finger-like villi which cover the surface of the small intestine.
Redrawn with permission from J.B. Taylor and P. D. Kennewell, *Modern medicinal chemistry*. New York: Ellis Horwood, 1993.

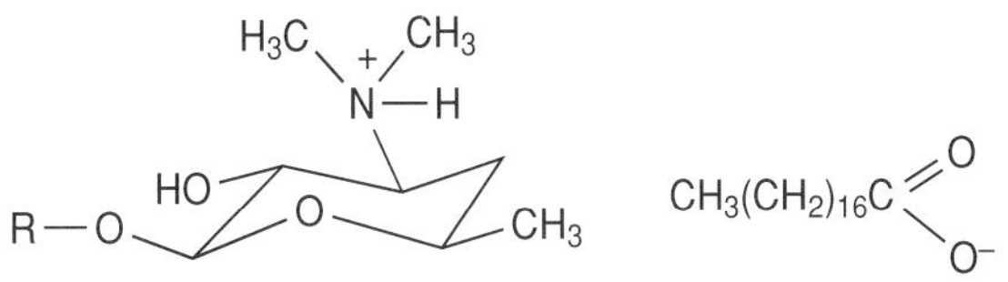

Figure 7 The stearate salt of erythromycin

RS•C

One solution which has been adopted is to make a derivative of erythromycin – the stearate salt (*Fig 7*). This salt is so insoluble that it passes through the acidic conditions of the stomach unaltered. On entering the small intestine, where the pH is higher, (more alkaline) the stearate salt releases free erythromycin base, which is stable at this pH, and is absorbed through the gut wall and into the blood plasma.

2. Children

Erythromycin is used regularly with young children who are prone to respiratory tract infections. The typical dose of erythromycin is 250 mg which, along with other constituents, makes for a large tablet. Patients of this age do not normally like taking tablets or capsules, and a liquid formulation (measured out by the spoonful) is preferred. The very bitter taste of erythromycin is such that children will not take the medicine and it cannot easily be masked with sweeteners such as sugar.

One solution is to make an ethyl succinate ester derivative of the erythromycin (*Fig 8*). This is extremely insoluble, so much so that it cannot be tasted. The insoluble ethyl succinate is taken mixed with water, forming a smooth suspension. This insolubility protects it from reaction with stomach acid. In the higher pH conditions of the small intestine, some of the derivative breaks down to release free erythromycin which can be absorbed. Some of the ethyl succinate derivative is absorbed unchanged through the gut wall and into the blood plasma. Here it is broken down by various non-specific esterase enzymes to free erythromycin.

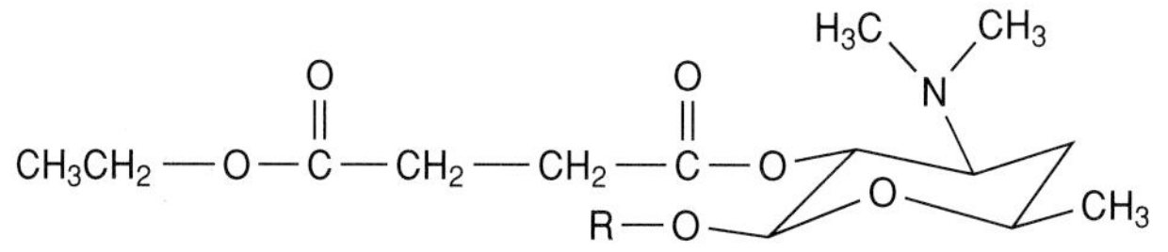

Figure 8 The ethyl succinate ester of erythromycin

Question 7

Mark on a copy of the structure of the ethyl succinate ester of erythromycin the ethyl succinate ester linkage. What small molecule is involved in the formation and breakdown of the ester?

3. The very sick

Very sick patients cannot always take medicines by mouth so they must take them intravenously – by injection or drip. Medicines taken in this way must be completely soluble, as small, insoluble particles could block capillaries, leading to haemorrhages or strokes. It is more convenient if solutions for intravenous use are fairly concentrated. This means that for injections, only small quantities need to be used. If the solution is to be used for an intravenous drip, it can be diluted in an appropriate solution. Erythromycin is not very soluble at pH 7 but it can be made soluble by combining it with lactobionic acid (*Fig 9*).

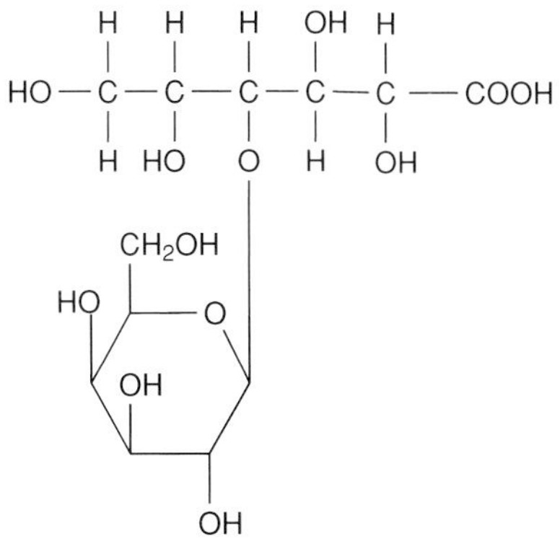

Figure 9 Lactobionic acid

Once in the blood plasma, the combination is converted back to free erythromycin. This is now much less concentrated (being dissolved in the whole blood volume, not just the injection solution) so that no precipitation occurs.

Question 8

What type of interactions govern the water-solubility of organic compounds? Explain why the stearate salt of erythromycin is less soluble than erythromycin itself and why lactobionic acid is very soluble in water.

Boxes 8 and 9 contain two other formulation case studies.

RS•C

Formulating paracetamol tablets   **Box 8**

Paracetamol is available over the counter as a mild pain reliever which also relieves fever. One issue when formulating paracetamol is that the dose required to have an effect (the therapeutic dose) is close to the effect at which poisoning occurs. This means that it is easy to overdose on paracetamol either accidentally or deliberately. Moreover, many proprietary remedies – cold cures *etc* – contain paracetamol, so that it would be easy to take a headache pill and a cold remedy as well, without realising that both contained the same drug. Overdoses of paracetamol cause irreversible liver damage but symptoms do not appear for 48 hours, by which time the damage is done.

An antidote to paracetamol poisoning (the amino acid methionine) is available. It is effective provided it is taken quickly. Because of these factors it has been suggested that paracetamol should be formulated with methionine added so that poisoning by an overdose becomes impossible. One brand (Paradote) is now available with this formulation.

In the liver, paracetamol is excreted partly by oxidation to a toxic compound called N-acetyl-*p*-benzoquinonimine (NAPQI) which then reacts with a compound called glutathione. In an overdose, the liver's reserves of glutathione become depleted and the toxic NAPQI cannot be deactivated. Glutathione is made from the amino acid methionine. So the extra methionine replenishes the liver's reserves of glutathione allowing the toxic NAPQI to be removed.

The structures of paracetamol and methionine

Question 9

Perform a SWOT analysis (Strengths, Weaknesses, Opportunities and Threats) on the issue of formulating paracetamol with methionine. What are your conclusions? Should paracetamol be available like this? Should it *only* be available like this?

Formulating effervescent tablets   Box 9

Some tablets are formulated in effervescent form – they fizz as they dissolve in water. Some people prefer to take medicines as a drink rather than as a tablet and the effervescence helps the tablet dissolve to give a clear solution. The fizzing is caused by carbon dioxide gas produced from the reaction of citric acid and sodium hydrogencarbonate:

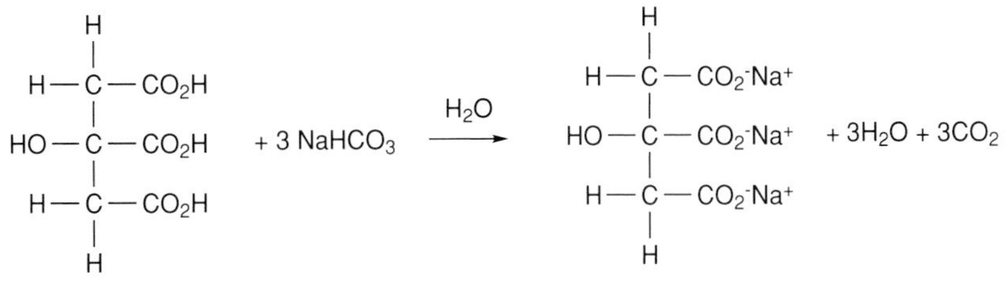

The reaction occurs only in the presence of water. This can cause manufacturing problems as the product must be manufactured in a humidity-controlled room. The selection of a lubricant for the tablet can also be a problem. The usual waxy, hydrophobic materials such as magnesium stearate leave a scum on the water surface, so a soluble lubricant such as polyethylene glycol must be used.

Question 10

Use the equation above to calculate the ratio by mass in which citric acid and sodium hydrogencarbonate react.

A tablet actually contains 105 mg of citric acid and enough sodium hydrogencarbonate to react with it all. What volume of carbon dioxide will it produce? (Assume 1 mole of any gas occupies 24 dm³ at room temperature and pressure.)

RS•C

Formulating agrochemicals

Agrochemicals are chemicals applied to crops, and include herbicides, fungicides and insecticides. Many of the issues in their formulation are similar to those in the pharmaceutical field. The job of the agrochemical formulator may be defined as the conversion of a **crude active ingredient** into a **convenient to use** preparation which has the **optimum biological activity**, is **stable** to physical and chemical degradation, and presents a **low hazard** in both handling and application.

Agrochemicals tend to be sprayed onto crops, and one of the formulation challenges is to get the active ingredient into a form suitable for spraying. Frequently the problem is to produce a product which can be diluted with water to be applied as a solution, emulsion or suspension.

Question 11

Explain the difference between a solution, an emulsion and a suspension.

When formulating agrochemicals, the considerations include:

▼ the biological activity and mode of action of the active ingredient

  – does it affect plants, insects, fungi *etc* ?

  – how is it absorbed into the plant and target organism?;

▼ the physical and chemical properties of the active ingredient

  – is it soluble in water or non-polar solvents?

  – is it a solid or a liquid?

  – how volatile is it?;

▼ the required method of application;

▼ shelf life – stability to heat, humidity, light, *etc*;

▼ ease of use;

▼ safety and environmental considerations;

▼ the costs of the raw materials and of the manufacturing process;

▼ commercial preferences; and

▼ ease of dispersion and/or dilution in various types of water.

Question 12

Which of these considerations will also affect the pharmaceutical formulation process? Are there any considerations which will affect a pharmaceutical formulator which will not affect the agrochemical one? Will any of the considerations vary significantly in importance between the two processes?

Microencapsulation

This is a formulation technique widely used by Zeneca, a large UK manufacturer of agrochemicals. It involves delivering the product in liquid droplets (typically 1–20 micrometres in diameter) dispersed in water. Each droplet is surrounded by a polymeric membrane through which the active ingredient diffuses.

This system may seem complex but has a number of advantages.

▼   A significant reduction in toxicity to humans because of the slow release of the active ingredient compared with neat active ingredient.

▼   Protection of the active ingredient from degradation by light and moisture.

▼   The rate of release of the product can be controlled by varying the thickness, permeability and amount of crosslinking of the membrane and the size of the droplet.

The formulation of the pyrethroid insecticide Lambda-cyhalothrin (*Fig 10*) is a good example of the technique. Lambda-Cyhalothrin is used to control insects such as cockroaches, mosquitoes and tsetse fly which are vectors (carriers of) a number of serious diseases especially in developing countries (*Table 1*).

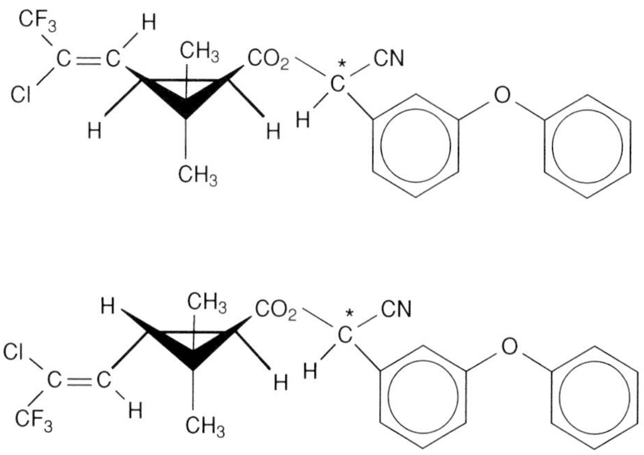

Figure 10 Lambda-Cyhalothrin is a mixture of the two isomers shown

Question 13

Lambda–Cyhalothrin is in fact a mixture of two isomers whose strucures are given in *Fig 10*. Look at the structures at the carbon atom marked * and note the differences. What type of isomerism is involved here? How would you be able to distinguish between the two isomers?

Pest carrier	Disease transmitted	Cases per year	Number of people at risk
Cockroaches and flies	Dysentery	8 600 000	5 000 000 000
Mosquitoes	Malaria Filariasis Arboviruses	1 500 000	4 700 000 000
Reduviid bugs	Chagas	100 000+	65 000 000
Sandflies	Leishmaniasis	1 000+	500 000 000
Tsetse fly	Sleeping sickness	ca 1 000	ca 10 000 000

Table 1 Diseases carried by various insects

RS•C

The product is used in public health situations – *ie* sprayed in houses *etc* and needs to remain active and control insects for up to six months. It therefore needs to work on a variety of different surfaces, porous and non-porous; be resistant to ultraviolet light and be of low toxicity to humans and other mammals which might come into contact with it. One of the main modes of action of the microcapsules is by being picked up on the insect's body (*Fig 11*). The active ingredient then diffuses through the capsule's membrane, through the cuticle of the insect and on to the site of action. Other possibilities are ingestion of the capsule by the insect or the rupture of the capsule on the insect's body leading to gross release of the active ingredient.

Figure 11 Lambda-Cyhalothrin microcapsules picked up on the legs of German cockroaches (*Blatella Germanica*). Courtesy of Zeneca

Encapsulating the active ingredient

The polymer used to form the capsule is a poly(urea) made by polymerisation of di- and tri-isocyanates according to the reaction scheme in *Fig 12*.

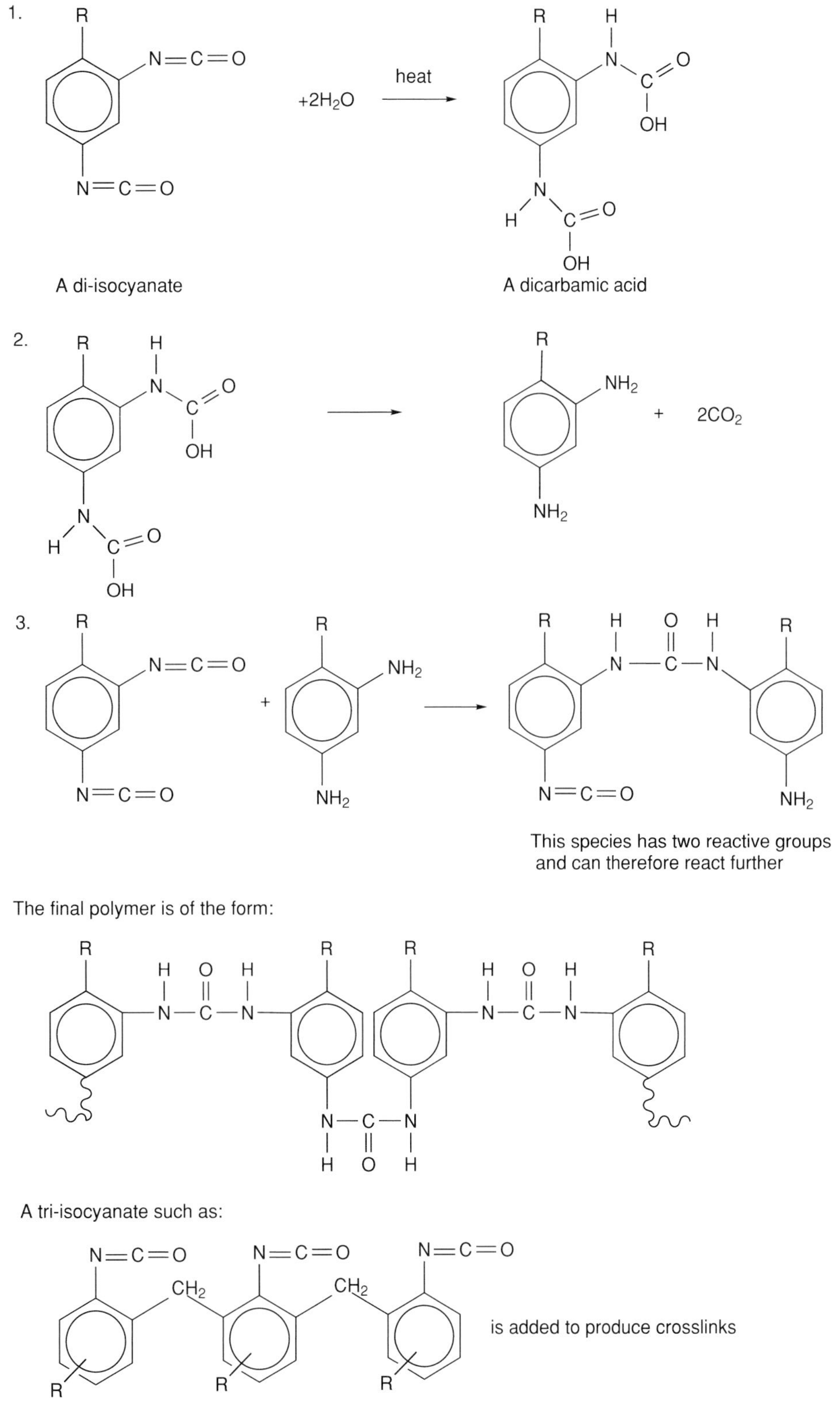

RS•C

Figure 12 The formation of a poly(urea)

RS•C

Some of the di-isocyanate reacts with water on heating to form first a dicarbamic acid and then a diamine. Diamines react with di-isocyanates to form the polyurea polymer. Some tri-isocyanates are added to the reaction mixture to produce cross-links.

Question 14

a)   Suggest how the properties of the crosslinked polymer might compare with the non-crosslinked version.

b)   Classify each of the steps in the reaction scheme as addition, elimination or substitution.

At first sight, coating droplets of oil 1–20 micrometres in diameter with polymer seems an impossible task. It is done by first making a solution of the active ingredient (Lambda-cyhalothrin) in a solvent, called the oil phase, which does not mix with water. The di-isocyanate monomer is added to this solution. The solution is stable at room temperature for several hours. The mixture is added to water containing an emulsifier and mixed vigorously to form an emulsion of oil droplets in water, stabilised by the emulsifier. The size of the droplets can be controlled by regulating the amount of mixing and the concentration of emulsifier – more mixing and more emulsifier both giving smaller drops. Polymerisation is then initiated by heating the emulsion to
50–60 °C. The polymerisation reaction occurs at the oil-water interface and can be stopped by adding ammonia or water soluble amines to quench the reaction by "mopping up" unreacted isocyanates.

Question 15

Suggest what features a molecule of emulsifier would be expected to have so that it can promote mixing between droplets of oil and water.

Question 16

Explain why the polymerisation reaction takes place only on the surface of the droplet and not throughout it.

Question 17

Write an equation for the reaction between a di-isocyanate and ammonia. How does it quench the polymerisation reaction? Hint: look at reaction 3.

Question 18

How would you use ammonia to regulate the thickness of the capsule?

We now have an emulsion of polymer-encapsulated micro droplets in water. This concentrate can be sold for dilution with more water to form a mixture suitable for spraying.

Answers to questions

1.

a) The amide linkage

$$\downarrow H_2O \:/\: H^+$$

b)

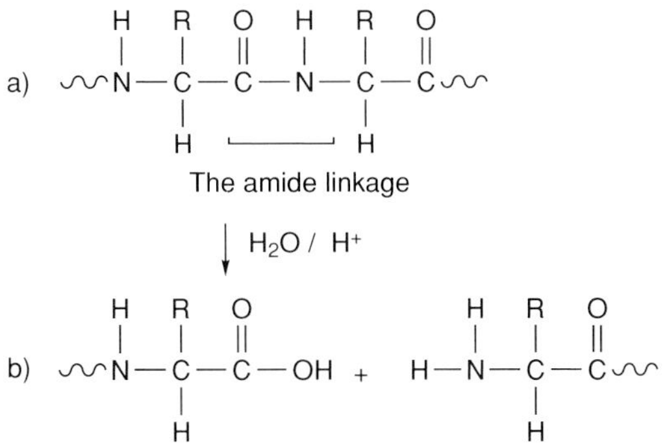

2.   Answer depends on method chosen.

3.   Factors include: state of subdivision of active ingredient, coating of the tablet, proportion of filler and active ingredient *etc.*

4.

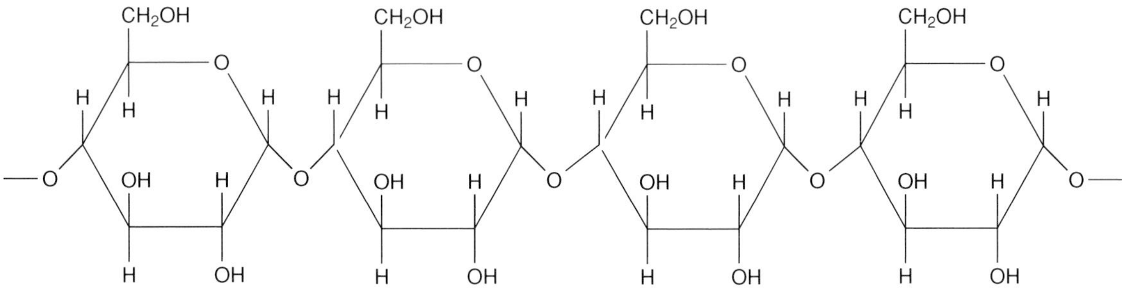

a)   Starch has many hydrogen bonding (–OH) groups in the molecule, so it is likely to absorb water, which can form bonds with these groups. This may cause the structure to swell.

b)   Stearic acid has a long, non-polar hydrocarbon chain which is unable to

RS•C

form hydrogen bonds with water.

5. Phenylmethanol is not a phenol because the –OH group is not attached directly to the benzene ring.

6. a)

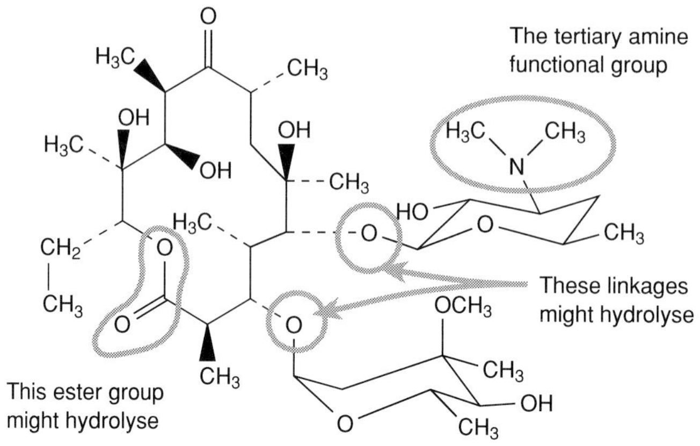

b) The ester functional group might be expected to be hydrolysed in acid conditions as might the R–O–R linkages marked.

c)

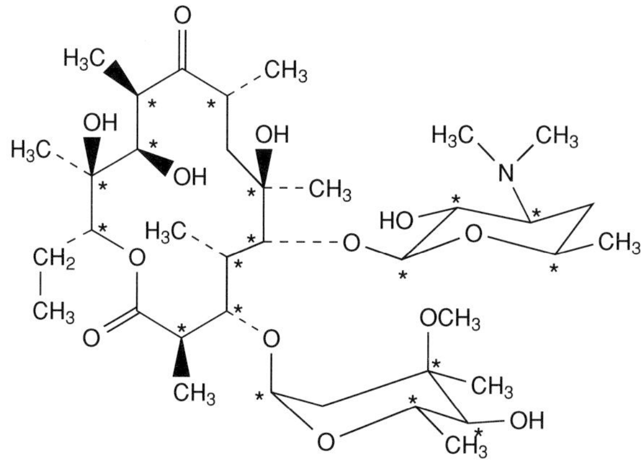

7.

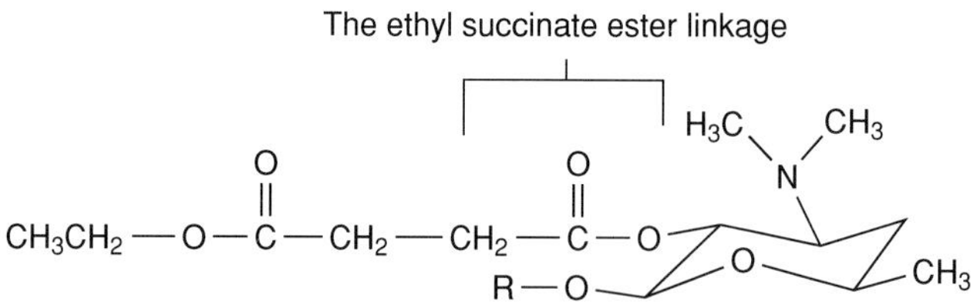

The ethyl succinate ester linkage

A water molecule is eliminated when the ester is formed and added on when the ester is hydrolysed.

8. Water soluble compounds form intermolecular interactions (hydrogen bonds or dipole-dipole bonds) with water. Hydrogen bonds form via –OH and –NH groups but not via –CH groups. The stearate salt has a long hydrocarbon chain which cannot hydrogen bond to water and therefore is less soluble than the parent erythromycin. Lactobionic acid has several –OH groups and is therefore soluble.

9. Factors which might be suggested include: safety, extra cost, public image of the company, marketing considerations and quasi-philosophical ideas such as whether "nannying" in this way is a good thing.

10. 210 g citric acid react with 252 g of sodium hydrogencarbonate. So the reacting ratio is 1: 1.2.
36 cm^3 carbon dioxide.

11. In a solution the two (or more) components are mixed at the molecular level. An emulsion is a mixture of small droplets (10^{-7}–10^{-9} m in diameter) of one liquid dispersed in another. It will not normally separate out. In a suspension, larger particles (often of a solid) are dispersed in a liquid. The suspension will gradually settle out of its own accord.

12. All of them (to a greater or lesser extent). Toxicity to humans will be of even greater importance to the pharmaceutical formulator. The effects on plants, insects and fungi will be less important to the pharmaceutical formulator. A variety of acceptable answers (with supporting arguments) is possible.

13. This carbon is chiral (is attached to four different groups). The molecule will exist as a pair of optical isomers. Each isomer will rotate the plane of polarisation of polarised light in the opposite sense.

14. a) The crosslinked polymer is likely to be more rigid and will not soften as easily on heating. This is because the molecular chains cannot slide past one another.

    b) Step 1: addition

       Step 2: elimination.

       Step 3: addition.

15. An emulsifier molecule would be "tadpole shaped" and have a water soluble (polar or ionic) "head" and a non-polar "tail" which is soluble in non-aqueous solvents.

16. Polymerisation requires water (for step 1) which is only present at the surface of the drop.

17.

RS•C

This stops polymerisation by converting the isocyanates into molecules which cannot react further.

18.   Adding more ammonia would shorten the polymer chains and make the capsule thinner.

RS•C

Developing a fungicide, azoxystrobin

The world wide market for agricultural fungicides (compounds used to treat fungal diseases in plants) is around £4 billion per year. This study looks at the development by Zeneca of a new plant fungicide, azoxystrobin, formerly code named ICIA5504 , and the processes which it has to go through from the initial discovery to being on sale – a process that can take up to 15 years. Incidentally the code name reflects the fact that the development of ICIA5504 began before the de-merger of ICI and Zeneca in 1993. The story is typical of the development of other agrochemicals (chemicals used by farmers to treat crops). *Figure 1* summarises the main parts of that process diagramatically.

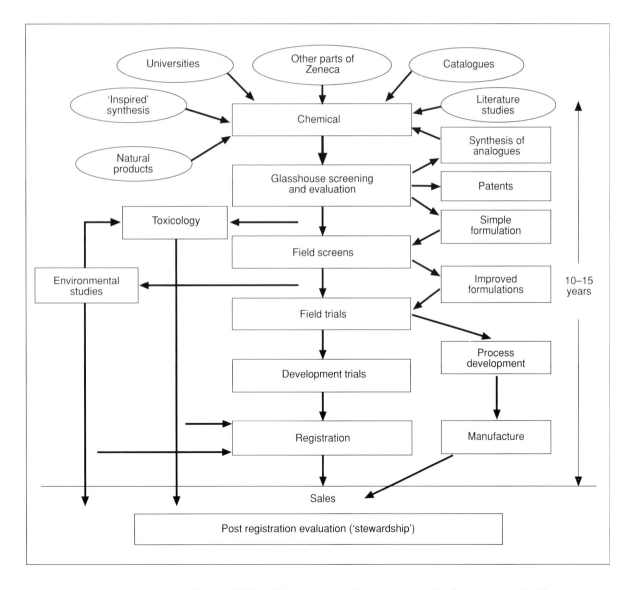

Figure 1 The discovery and development of a new pesticide

RS•C

Table 1 Many people from several disciplines are involved over a period of years in the development of an agrochemical such as azoxystrobin

Year	1	2	3	4	5	6	7	8	9
Research chemist	Selection of compound and small scale synthesis (100 g)								PRODUCT LAUNCH
Economist		Feasibility studies				Choice of manufacturing site		Construction of plant	
Toxicologist	Acute toxicology tests		Studies on production samples						
Environmental scientist			Determination of fate in the environment. Preliminary studies on ecotoxicity	→	→	→	→		
Biologist	Screening for activity		Tests in the field	→	→	→	→		
Patent officer	Patent application						More patent applications		
Process development chemist		Pilot plant for larger scale production (100 kg) for testing		Development of large scale synthesis	→	→	→		
Registration officer					Application for registration				
Formulation chemist				How to apply the required dose?	Stability tests. Packaging				
Marketing specialist		Compare with current market leader or is this a new market?				How and where to launch the product?			

RS•C

Development of a fungicide or other agrochemical is carried out by multidisciplinary teams. Table 1 shows some of the disciplines involved, what each one does and when, although details and time scale will vary from product to product. In this study we look at the discovery and environmental testing of azoxystrobin.

Discovery

As indicated in *Fig 1*, the initial discovery of compounds with particular types of activity comes from a variety of sources. In a typical year, Zeneca might test 100 000 compounds for activity of which fewer than 500 might be selected for further testing and 1 or 2 might eventually be developed.

In the case of azoxystrobin, a family of natural products, the strobilurins, oudemansins and myxothiazols (all related in structure to β-methoxyacrylic acid – structures 1–4) were found to have fungicidal activity. This was noted in 1981 when an ICI researcher read an account of them in a German research paper.

Note. β-methoxyacrylic acid's systematic name is 3-methoxypropenoic acid but the systematic names of the other compounds are too long for everyday use. In this case study we shall use the trivial names by which they are normally known.

Despite having fungicidal activity, these compounds were found in fungi growing on decayed wood. It is not yet clear how it is that these fungi are not affected by the fungicidal compounds they themselves synthesise. However, the fungicides seem to help the parent organism to compete with other fungi. The fungicidal compounds have been found to work by inhibiting electron transport within the cells of fungi.

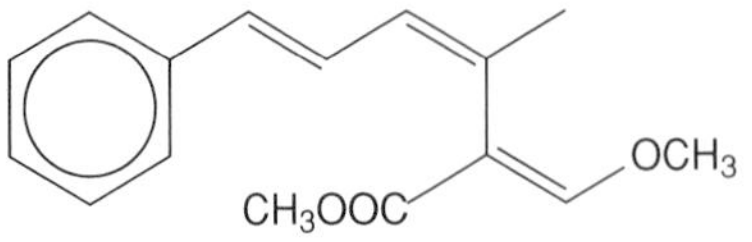

RS•C

(1) Strobilurin A

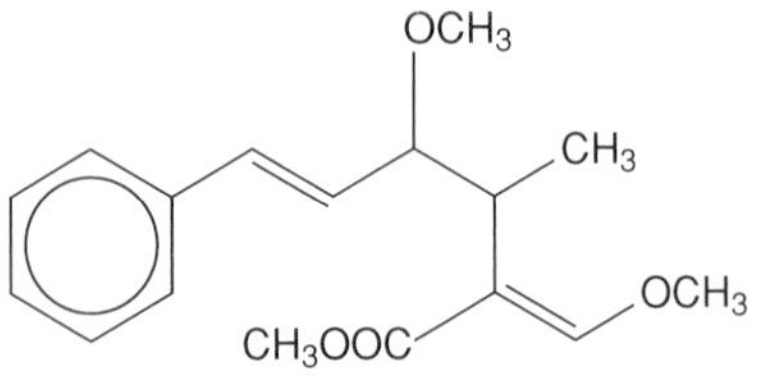

(2) Oudemansin A

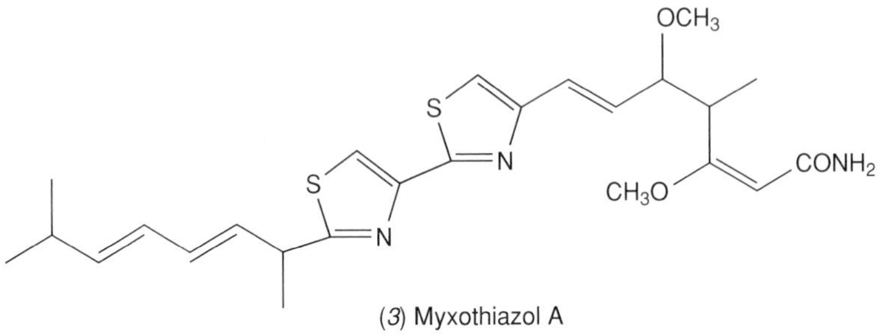

(3) Myxothiazol A

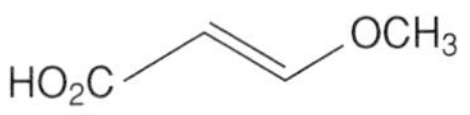

(4) β-methoxyacrylic acid

Fig 2 β-Methoxyacrylic acid and related compounds

Initially the fungicidal activity of these natural products was confirmed by glasshouse tests.

RS•C

A structural problem   Box 1

The simplest compound of the series was strobilurin A and it was considered important to establish its fungicidal activity. To do so, a method of making it in the laboratory had to be devised. The structure of strobilurin A had been reported in the literature as (5). Two syntheses were devised to make the compound with this structure but it was found not to be the same as strobilurin A. Careful analysis of the spectroscopic data indicated that strobilurin A had the structure (1) and a synthesis was devised which led to the correct compound – see below.

(5) The wrongly-reported structure of strobilurin A

The synthesis of strobilurin A used to confirm its structure

Question 1

Look at the structure of strobilurin A and also the proposed but incorrect one. Can you see how they are related? Making models might help if you have access to a molecular modelling kit.

Strobilurin A (1) which is structurally the simplest compound, proved relatively ineffective in glasshouse trials as it broke down quickly in light and was also quite volatile. This meant that it did not remain on the leaf for long after spraying. It did, however, show activity against fungi growing on agar in petri dishes in low light conditions.

So a number of derivatives (related compounds) of strobilurin A were synthesised. These retained the β-methoxyacrylate unit which was believed to be responsible for the fungicidal activity of the compounds. These included compounds (6), (7), and (8) (Fig 3).

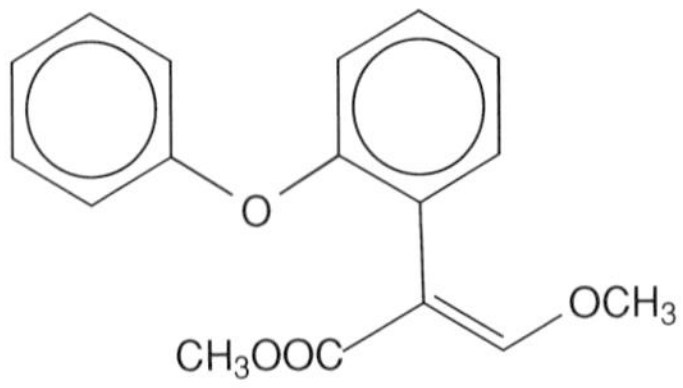

RS•C

(6)

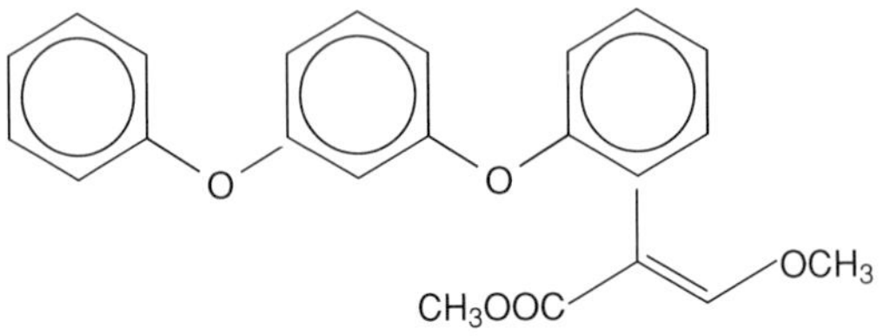

(7)

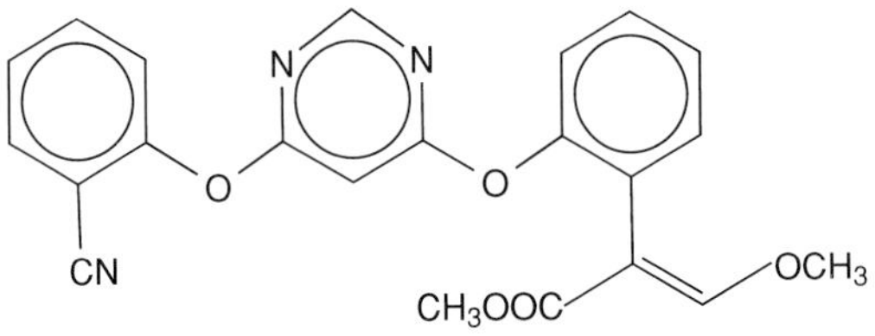

(8)

Figure 3 Compounds synthesised during the development of azoxystrobin

Compound (6), containing a diphenyl ether unit turned out to be less volatile and more photochemically stable than strobilurin A. It moved systemically through the plant, *ie* it was transported through the plant tissue – an important advantage. It showed good fungicidal activity but it also damaged the plants in trials.

Addition of a further aromatic ring gave compound (7), which had even better fungicidal activity but was too insoluble in water to move systemically.

Incorporation of electronegative nitrogen atoms to form a heterocyclic aromatic ring improved the water solubility and eventually azoxystrobin (8) was produced. This had a solubility in water of around 10 mg dm^{-3}, and, in a laboratory test, it took 20–35 hours for half of it to decay in light equivalent to a bright summer's day. It was effective against a range of fungi. In particular, it controlled both of the two key fungi which affect vines and rice. Previously, two different fungicides had to be applied to control these.

Question 2
Can you explain why incorporating nitrogen atoms into the molecule improved its water solubility? What type of intermolecular forces are involved?

A further important point is that azoxystrobin has a relatively low toxicity. Toxicity is expressed as an LD$_{50}$ (lethal dose, 50%) – the dose (in grams of test substance per kilogram of body mass) required to kill half of a population of test animals. The *lower* the LD$_{50}$, the *more* toxic the substance. Azoxystrobin's LD$_{50}$ for oral administration to rats is 5 g kg^{-1}. For comparison, the LD$_{50}$ of aspirin is 1.5 g kg^{-1} and the LD$_{50}$ of caffeine is 0.13 g kg^{-1}. So azoxystrobin is less toxic than either of these everyday compounds.

At this stage, the Zeneca management had to take a decision whether to proceed with azoxystrobin or not. To proceed would mean spending several million pounds on toxicity and environmental tests and, at the same time, devising a suitable method to make it on a large scale. These projects had to run in parallel for time reasons – to run them one after the other would lead to an unacceptable delay in getting the product on the market. However, this involved the risk that spending on the synthetic method would be wasted if the compound failed in the other trials.

Toxicology

Preliminary studies must be done to measure the toxicity of a compound (when ingested by mouth, through the skin or by inhalation) for legal reasons and so that appropriate precautions can be taken for use, handling and transport during development. A compound might be discarded at this stage because it could be too toxic to develop and use. There is also an ethical decision to be made – should Zeneca sell such a compound? Later on, toxicology tests have to be repeated using material made by the production process as different methods of synthesis might generate different impurities which may themselves be toxic.

Environmental fate and effects

For all agrochemicals, two key questions must be answered before the substance can be registered and sold.

1. What happens to the chemical after it has been applied?

2. What effect does the chemical (and its breakdown products) have on living things (plants and animals) exposed to it?

Question 1 relates largely to the chemistry of the compound – its solubility and stability. It involves the rate of breakdown when exposed to soil, water, light and in plants and animals. It also involves identifying the breakdown products and their properties, ie toxicity, water solubility, do they bind to soil? Measurements are made of the concentration of the compound and its breakdown products (called "degradates", or "metabolites" if they are the result of biological degradation) in soil, water, the crop itself and other plants and animals which might be exposed to it in a variety of ways. This is done first in laboratory or glasshouse trials and then by trials in the field – usually in a number of different places across the world and over a number of years so that a variety of different conditions is experienced. At Zeneca, samples of plant material, soil and water are brought to the Jealott's Hill Research Station in Berkshire for measurements of residue levels.

Question 2 relates to the biological effects of the compound and its breakdown products and their effects on a variety of living things from microbes to mammals.

Chemical testing

Much of the chemical testing is concerned with measuring levels of residues of the applied chemical and its degradates and metabolites in a variety of samples, routinely down to levels of 0.01 part per million (ppm). This is done by grinding the samples, extracting the residues into a suitable solvent and separating them by gas chromatography (GC) or high performance liquid chromatography (HPLC) with a suitable detector, often a mass spectrometer.

Note. A level of 0.01 ppm is equivalent to finding one sheet of paper in a pile of paper the height of Mount Everest.

RS•C

High performance liquid chromatography (HPLC)   **Box 2**

High performance liquid chromatography (HPLC) is one of the most important modern chromatographic separation techniques. A solvent mixture (the mobile phase) is forced at high pressure through a powdered solid stationary phase (often silica). The mixture to be separated is injected into the mobile phase and is separated on the column. The detector is often a mass spectrometer which can identify components from their relative molecular masses and fragmentation patterns. Very efficient separations are possible with HPLC – even pairs of optical isomers can be distinguished with an appropriate stationary phase.

As well as chemical testing, biological tests are carried out on the effects of agrochemicals on non-target plants, soil micro-organisms, earthworms, insects, birds, fish, aquatic invertebrates, algae and mammals both in the field and the laboratory. Non-target plants are those which might be treated by accident (by spray drift, for example) as opposed to the crop to which the fungicide is deliberately being applied.

Studies include acute tests (which measure lethality) and chronic (long term) studies which measure effects on growth and reproduction of organisms exposed to the test chemical. The potential of the test chemical to accumulate in the food chain is also measured.

For example, tests of the effect of azoxystrobin on the aquatic environment included acute and chronic studies on fish species (including trout, carp and bluegill sunfish), invertebrates (such as *Daphnia*) and algae. The concentrations at which effects were observed were then compared with the concentrations which were predicted to occur in the environment, to see if adverse effects were likely. Toxicity tests were also carried out on a variety of invertebrates including insects, snails, worms and zooplankton. Bioconcentration factors may also be measured for some products. These are indications of how much a substance is concentrated in different organisms. For example, a fish might be found to have 100 times the concentration of a test chemical in its body compared with the water in which it was living.

Tests were carried out in both laboratory and field and the results indicated that azoxystrobin would pose a negligible risk to the aquatic environment.

RS•C

Assessing risk   **Box 3**

Much of the development work on agrochemicals is involved with the assessment of the risk involved in exposure to certain chemicals. The risk of exposure is related to two factors:

▼   the inherent toxicity of the chemical itself which can be expressed by its LD_{50}; and

▼   the level of exposure to the chemical.

So even a highly toxic compound presents very little risk if the exposure is very low and conversely even a relatively innocuous chemical, such as caffeine, can kill if the dose is high enough.

For residues in the diet, an Acceptable Daily Intake (ADI) is used to estimate the amount of a residue that can be safely consumed over a whole lifetime. The ADI is derived from feeding trials on a variety of laboratory animals. These enable a no observed effect level (NOEL) to be found for the species which is most sensitive to a particular chemical. An uncertainty factor of 100 (1000 for chemicals which have shown any evidence of causing cancer) is then built in to give an ADI which is measured in mg of chemical per kg of body mass per day.

Radiolabelling

Radiolabelling can be useful in studying where the compound and its breakdown products appear in the environment. The pesticide can be synthesised so that radioactive atoms (usually ^{14}C) are incorporated in it. ^{14}C is a β-emitter so the fate of the compound and any of its breakdown products which contain ^{14}C can be traced by measuring the radioactivity emitted. That is, the location of the activity can be traced even if the chemical identity of the product is unknown. Radiolabelled compounds are expensive to prepare so once the radioactivity has been located and the compound it is in has been identified, further work can proceed using unlabelled material. ^{14}C-containing compounds can also be detected by mass spectrometry because a fragment ion containing a ^{14}C atom has a mass two units greater than one containing ^{12}C.

RS•C

Synthesising ^{14}C labelled compounds   **Box 4**

^{14}C is made in a nuclear reactor by bombarding ^{14}N containing-compounds with neutrons. This is followed by ejection of a proton from the nitrogen nucleus to form ^{14}C.

$$^{14}_{7}N + ^{1}_{0}n \rightarrow ^{14}_{6}C + ^{1}_{1}p$$

The resulting $^{14}_{6}C$ decays back to $^{14}_{7}N$ by β-emission with a half-life of 5730 years. This long half-life means that there will be no significant loss of radioactivity caused by radioactive decay during the time scale of the experiments (less than 10 years).

The compounds which are bombarded with neutrons are beryllium nitride (Be_3N_2) or aluminium nitride (AlN), chosen for their high nitrogen content, stability to heat and radiation and lack of contamination with ^{12}C. The resulting beryllium or aluminium carbides are oxidised to give $^{14}CO_2$ and reacted with barium hydroxide to give barium carbonate ($Ba^{14}CO_3$) from which $^{14}CO_2$ can be produced. This is the starting material for all ^{14}C-labelled organic compounds.

The labelled carbon can be incorporated into organic compounds in a variety of ways.

1. Carboxylation via a Grignard reaction to give carboxylic acids labelled at the carboxyl carbon, for example:

2. Reduction to methanol by lithium aluminium hydride (lithium tetrahydridoaluminate(III)), for example:

$$^{14}CO_2 \xrightarrow{\text{LiAlH}_4} {}^{14}CH_3OH$$

3. Reaction with ammonia followed by reduction to sodium cyanide which can then be converted into nitriles:

$$Ba^{14}CO_3 + NH_3 \longrightarrow BaN^{14}CN \xrightarrow[\text{heat}]{\text{Na}} Na^{14}CN$$

$$R - Br + Na^{14}CN \longrightarrow R^{14}CN + NaBr$$

4. Reduction to barium carbide which can then be hydrolysed to give ethyne which can in turn be used to make benzene and hence a variety of aromatic compounds:

$$Ba^{14}CO_3 \xrightarrow{\text{Ba}} Ba^{14}C_2 \xrightarrow{\text{H}_2\text{O}} H^{14}C \equiv {}^{14}CH \longrightarrow$$

Question 3

How would you attempt to make the following ^{14}C isotopically labelled compounds?

a.   Methyl benzoate labelled on (i) the aromatic ring, (ii) the carboxyl carbon, (iii) the methyl group.

b.   Propanoic acid with the label on the carboxyl carbon.

c.   Phenylamine (aniline) with the ring labelled.

Radiolabelling trials have to be small scale for economic reasons because the labelled compounds are expensive to synthesise. Typically a trial might be done on a 1 m² plot of crop and require 25 mg of radiolabelled compound costing over £1000. A typical trial with a radiolabelled compound might go as follows (*Fig 4*).

   The labelled test compound is sprayed onto a crop which is then harvested, say one month later. The crop is ground up and shaken with an appropriate solvent in which the radiolabelled compounds are soluble. A portion of this is placed in a scintillation counter to determine the total radioactivity. The radioactive compounds are then partitioned between an aqueous and an organic solvent. The compounds in each layer are next separated by thin-layer chromatography (TLC) or high performance liquid chromatography (HPLC) to determine the number of breakdown products which contain radioactivity. The soil on which the crop has grown is treated in the same way. In the laboratory, air from above the growing crop may also be sampled to determine how much of the labelled carbon has degraded to carbon dioxide. In this way the fate of the original compound can be traced.

RS•C

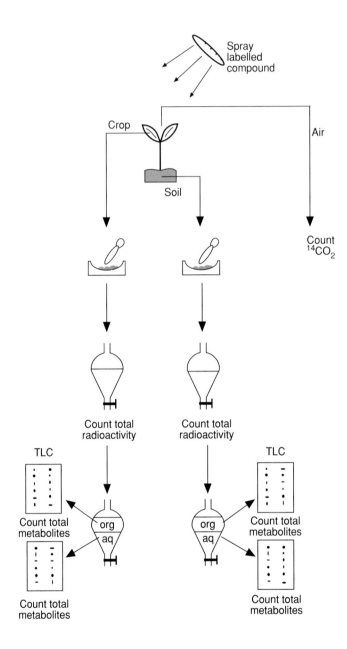

Figure 4 A radiolabelling trial

Interpreting information from radiolabelling

There are a number of issues to be considered when deciding where to label a compound. Take methyl benzoate, as a simple example. This could be labelled in the aromatic ring, the carboxylate carbon or the carbon of the methyl group (*Fig 5*).

Methyl benzoate

Ring labelled – any of the ring
poitions could be labelled

Carboxylate carbon
labelled

Methyl carbon
labelled

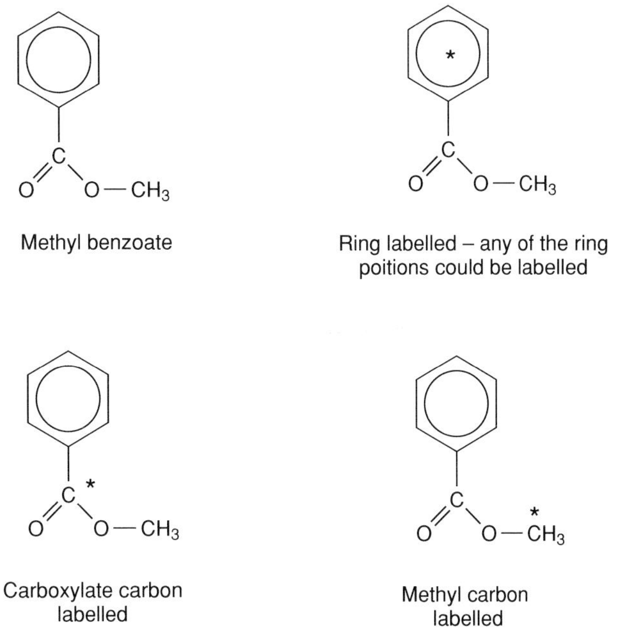

Figure 5 Possible labelling positions for methyl benzoate

However, labelling in this last position, the methyl group, would be of limited value in following the environmental fate of this compound as the methyl group will be easily lost by hydrolysis (quite likely in most environments).

Question 4
Write an equation for the hydrolysis of methyl benzoate in which the methyl group is lost. Under what sort of conditions would this take place?

So labelling in a more stable part of the molecule (such as the aromatic ring) is preferred by regulatory agencies. It is common to label more than one part of a molecule as is shown in the following (hypothetical) example with compound X (*Fig 6*).

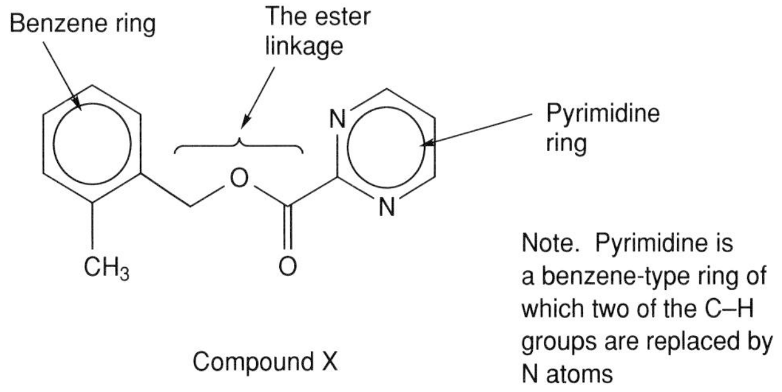

Benzene ring

The ester
linkage

Pyrimidine
ring

Note. Pyrimidine is
a benzene-type ring of
which two of the C–H
groups are replaced by
N atoms

Compound X

Figure 6 Radiolabelling of compound X

RS•C

Typically two radioactively-labelled samples of X would be synthesised, one with the label in the benzene ring and the other with the label in the pyrimidine ring. Let us call the benzene ring labelled compound B and the pyrimidine ring-labelled compound P. Two tests are then run, one treating a test system (crop, soil, water *etc*) with B and another an identical test system with P. Metabolites (compounds produced from X) M_1, M_2, M_3, and M_4 are separated by, say, thin layer chromatography. This is a technique similar to paper chromatography where the paper is replaced by a thin layer of a suitable stationary phase coated onto a glass or plastic backing. The β-radioactivity from the ^{14}C labels is counted by a technique known as autoradiography in which the chromatogram is sandwiched next to an imaging plate which is sensitive to radioactivity. A typical result might be as shown in *Fig 7*.

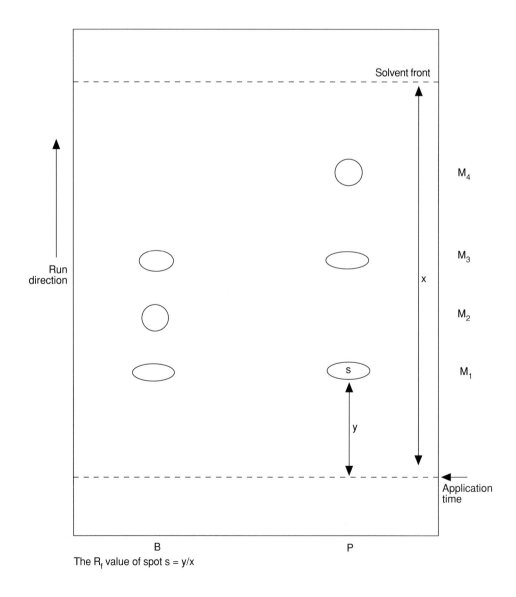

Figure 7 Autoradiogram of the metabolites of compound X

This indicates that M_1 contains both rings of the original compound X since radioactivity is counted from both B (benzene ring labelled) and P (pyrimidine ring labelled). M_2 does not show up in the test with P (pyrimidine ring only labelled) but does in the test with B (benzene ring only labelled). This suggests that M_2 contains the benzene ring only. By similar reasoning we can deduce that M_3 has both rings and M_4, the pyrimidine ring only.

Note that metabolites M_2 and M_4 are both present after both tests (the tests were identical) and are separated in the chromatograms. They contain no radioactivity because they are from the unlabelled halves of both B and P; only metabolites with a radioactive label show up on the autoradiogram.

These results suggest that the ester linkage in compound X might have been hydrolysed to form two fragments (M_2 and M_4) each containing one of the rings. M_1 and M_3, in which the two rings remain joined, could be a derivative of compound X in which one of the ring positions has been hydroxylated or in which the $-CH_3$ group has been oxidised to $-CH_2OH$ or $-CO_2H$ (*Fig 8*).

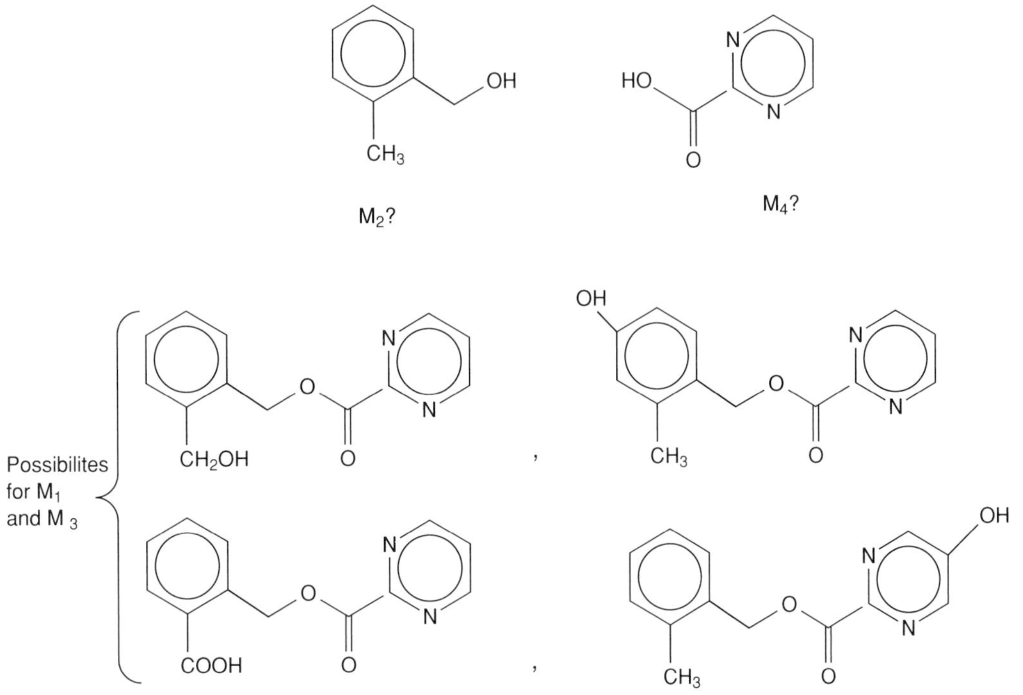

Figure 8 Possible metabolites of compound X

At this stage, of course, these are speculations which must be confirmed or rejected. For example, the proposed metabolites could be synthesised and TLCs run to find out if they give the same R_f values as M_1 to M_4. Alternatively, M_1 to M_4 could be isolated for identification by, for example, HPLC/MS (high performance liquid chromatography / mass spectrometry) in which each compound separated is fed directly into a mass spectrometer.

RS•C

The breakdown of azoxystrobin

Ultimately it should be possible to trace the breakdown of an agrochemical to carbon dioxide – the ultimate breakdown product of all carbon-containing compounds. For azoxystrobin, two breakdown mechanisms were found.

One was photolytic (ie light induced) and occurred on the surface of soil. The half life (ie the time for half the applied chemical to disappear) for azoxystrobin to break down to carbon dioxide by this process is about one to two weeks.

In the absence of light, azoxystrobin breaks down to carbon dioxide via microbial action with a half-life of about 80 days. This process results in the formation of an acid metabolite formed by hydrolysis of the ester group in azoxystrobin.

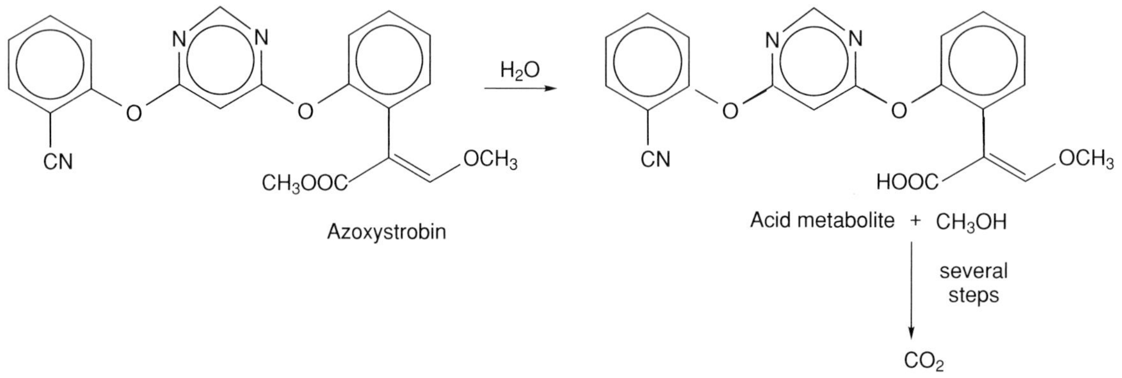

Figure 9 One step in the break down of azoxystrobin in the dark – hydrolysis of methyl ester to leave an acid

In the field, radiolabelling studies indicate that azoxystrobin breaks down with a half life of about 14 days and none of the acid metabolite is found. This suggests that photolysis is the main breakdown mechanism in the field. Similar results are found for the formulated product as for azoxystrobin alone. This rapid and complete breakdown is a very encouraging result as it indicates that there is little need to worry about long lived residues and breakdown products.

Mathematical modelling

Much of the data obtained from experimental work are used in mathematical models run on powerful computers which can help predict things such as the movement of agrochemical residues through ground water. This requires data about the chemical and its degradates (reactivity, solubility, volatility etc) as well as geological information about soil types and rainfall records over many years. The models themselves and the computers which run them are becoming increasingly sophisticated and it may soon become possible to predict residue levels in ground water in a particular place while a compound is still at the research stage. One Zeneca scientist has suggested that eventually such predictions could be made about hypothetical compounds which have not yet been made in the laboratory.

Registration

Before an agrochemical can be sold, it must fulfil the legal safety requirements of the country concerned, to ensure that it will not harm users, consumers of the crop or the environment. It must also be shown to be effective. Regulations vary from country to country and also depend on the circumstances of use. For example, a compound

RS•C

could be registered for all crops or for just a range of crops. The registration agency requires the results of a vast number of experiments and will specify what experiments must be done, and how. There is normally a dialogue between the Zeneca scientists and the specialists of the regulatory agency to ensure that the product is acceptable. Registrations can be cancelled or modified after being granted if unforeseen problems arise. The cost of the whole process required to register a new agrochemical in Europe is *ca* £50 million (1996 prices) of which 17% goes on efficacy, 50% on human safety and 33% on the environment. Products already on the market are kept under continuous review as they must be re-registered periodically.

Azoxystrobin will go on sale in different countries at different times as registrations are obtained. Different names will be used depending on the country and the use for which it is being marketed. In Europe it is called Amistar when sold for use on cereal crops. This product was launched in early 1997 and is also sold as Heritage when used in the US for use on golf courses.

RS•C

Answers to questions

1. See structures below. The difference is at the carbon atom marked x. In the correct structure, the methyl group and the chain containing the aromatic ring are *trans-* and in the incorrect structure, they are *cis-*. The two structures are isomers.

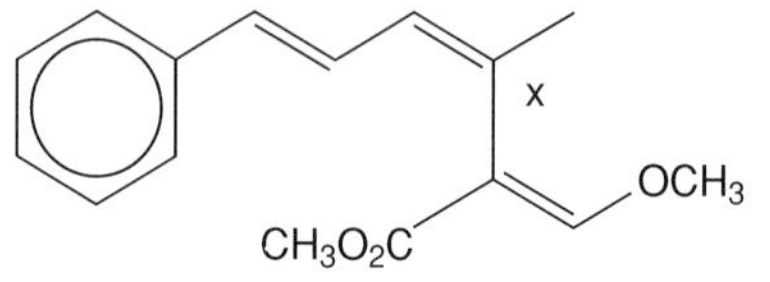

strobilurin A (correct structure)

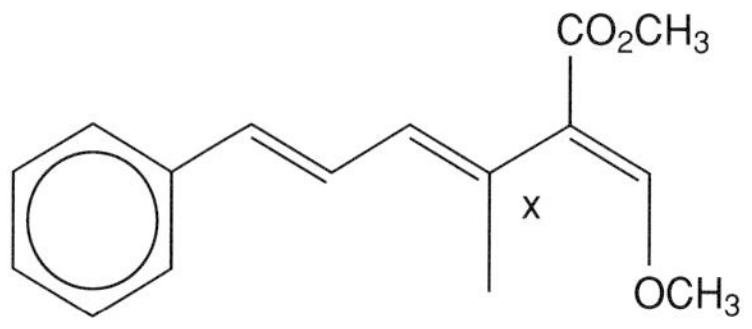

incorrect structure

2. Nitrogen atoms can participate in hydrogen bonding with water molecules.

3. a) (i)   Make labelled benzene as described in the box. Convert this to methylbenzene by a Friedel-Crafts reaction with chloromethane and an aluminium chloride catalyst.

        Oxidise the methyl group with acidified manganate(VII) ions to yield benzoic acid.

        Esterify with methanol.

    (ii) Make benzoic acid with the carboxyl carbon labelled as described in the box and then esterify with methanol.

    (iii) Esterify benzoic acid (or benzoyl chloride) with labelled methanol obtained as described in the box.

    b) Prepare labelled sodium cyanide as described in the box. React this with bromoethane to give propanenitrile with the label on the carbon of the –CN group. Acid hydrolysis will then yield propanoic acid labelled as required.

    c) Make labelled benzene as described in the box. Nitrate with concentrated nitric and sulfuric acid to yield nitrobenzene which can be reduced to phenylamine with tin and hydrochloric acid for example.

    Other synthetic routes are, of course, possible.

4.

Methyl benzoate $+ H_2O \longrightarrow$ Benzoic acid $+$ H–O–CH$_3$ Methanol

Water is required and the reaction is catalysed by both acids and bases. In the latter case, the salt of benzoic acid is formed rather than the acid itself.

RS•C

Scaling up a polymer manufacturing process

The Swindon-based company Raychem makes a variety of polymeric materials with properties tailored for specific applications, mostly in high-tech industries such as telecommunications. This case study looks at the introduction of a new material, called ACBS – adhesive cable blocking system. This is used for sealing bundles of telecommunication cables (*ie* wherever the cables are found) against water. In particular it looks at the problems which were encountered in making the material on a large scale in the factory (as opposed to on a laboratory scale) and how these were tackled.

The problem

There are a number of applications in which groups of electrical current-carrying wires are bunched together within an insulated sleeve (*Fig 1*). Examples include:

▼ telecommunication cables;

▼ power cables; and

▼ vehicle wiring harnesses (*ie* the bunches of wires which feed the instruments, lights, radio *etc* within cars).

Insulated outer sleeve

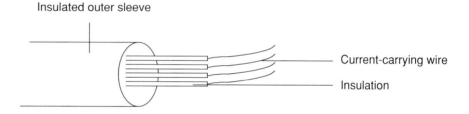

Current-carrying wire

Insulation

Figure 1 A typical bunch of cables

Block of polymer material

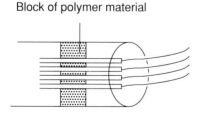

Figure 2 A cable block

It is important to prevent moisture getting inside the outer sleeve as this can lead to insulation breakdown, poor connections and other electrical failures. To prevent water getting in, the cable is blocked at various points. This involves sealing the cable with a filler material. It is usually applied to the cable as a liquid which fills the spaces between the inner wires. This then solidifies or thickens forming a permanent plug of sealant (*Fig 2*). This thickening may be brought about by applying the liquid hot and allowing it to cool and solidify, or by a chemical reaction called curing.

RS•C

Curing often involves the formation of crosslinks between polymer chains.

Cable blocking is frequently neccessary when a cable is repaired or a new length spliced on. This might occur at the bottom of a muddy trench. So the system must be easy to apply in less than ideal conditions by personnel without special chemical skills and knowledge.

Until recently, a material which could block a bunch of up to 200 pairs of cables within an outer sleeve was satisfactory. However, with the growth of communications, there is increasing demand for a product which will block up to 1000 pairs. This is more difficult as the gaps between the wires are smaller and it is harder for the blocking material to effectively penetrate in between them.

Some telecommunication cables are internally pressurised to prevent water leaking in so airtight blocking is essential to retain this pressure.

The market

At first sight, cable blocking may seem a small market but, to take just one application, there are 30 million cars made every year in Europe each of which has a wiring harness with some 30 splices. The overall value of the market for this application is around $50 million per year (1996 prices).

Research and development

Prior to the introduction of ACBS, Raychem already sold a cable blocking product called DWBS – dual water blocking system. This was based on a polyamide hot melt system in which a solid polyamide is applied to the cable and heated with a hot air gun so that it melts and the liquid penetrates between the individual wires of a bunch of cables. On cooling it solidifies and produces a seal. The product was actually applied to the cable by strapping a packet (made of a plastic net) of solid polymer around the cable splice and placing a sleeve of heat shrink polymer around it. This polymer shrinks on heating and this holds the packet of sealing compound in place as it is heated. On heating, the polymer sealant melts and moves through the net into the cable. At the same time, the gap in the cable's outer sleeve is sealed by the heat shrink polymer. (*Fig 3*).

Question 1
Polyamide plastics, such as Nylon, are made from two monomers, a diacid dichloride and a diamine. Draw structural formulae to represent each of these monomers and show how they react together to form an amide linkage. What feature of the monomers allows a polymer to form?

The DWBS system has the disadvantage that a high temperature is needed to melt the polyamide with a consequent risk of damage to the cables and their insulation. Also, the molten polymer is quite viscous, making it hard for it to penetrate effectively between the individual strands of wire. This means that it is not suitable for cables containing more than 200 pairs of wires.

A new product was required to overcome these disadvantages and meet the new requirement for blocking 1000-pair cables. In fact a suitable material was already known to the Raychem Research and Development team, who had tried it, without success, for two other applications. It is now the basis of a new cable blocking product called ACBS – adhesive cable blocking system. The material, on which ACBS was to be based, had a unique property known as supercooling. Supercooling is illustrated in *Fig 4*.

RS•C

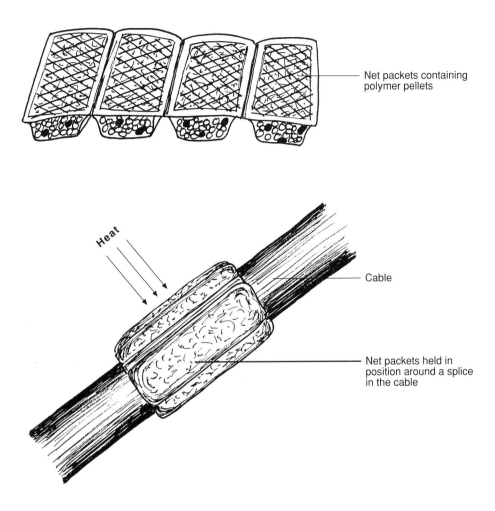

Net packets containing
polymer pellets

Heat

Cable

Net packets held in
position around a splice
in the cable

Figure 3 The application process and the net packet

RS•C

Heat shrink polymers   **Box 1**

Heat shrink polymers are another Raychem product. They are made from a
conventional poly(ethene) thermoplastic material. This is moulded into the
required shape (in this case a tube of the same diameter as, or slightly smaller
than, the cable it will be used on) and irradiated with β-radiation see below (*a*).
This forms crosslinks between the poly(ethene) chains. The tube is then heated
to soften it and re-moulded into a tube of larger diameter (*b*). This is then
cooled and retains its new larger diameter (*c*). In this situation the crosslinks
are stretched and, on re-heating, they pull the tube back into its smaller
diameter configuration (*d*). So the material shrinks on heating. It acts as though
it has remembered its original shape.

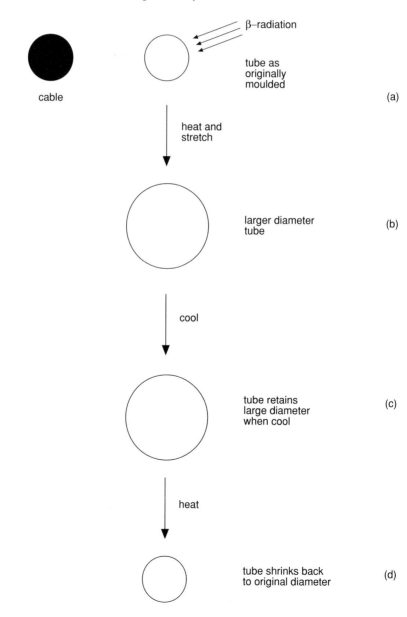

Cross-sectional views of the manufacture of heat shrink tubing

RS•C

1. Conventional blocking materials

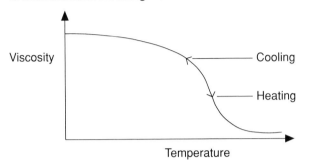

A conventional material's viscosity changes in the same way both on heating up and cooling down

2. Supercooling materials

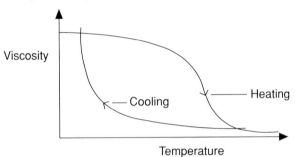

ACBS's viscosity remains low to a much lower temperature as it cools

Figure 4 The ACBS displays supercooling

When ACBS is heated, its viscosity slowly decreases – *ie* it gets thinner and flows better. However, on cooling, it remains as a thin, easily flowing liquid to a much lower temperature. This allows the material to flow into the voids between wires very easily.

ACBS is based on a polyurethane, made by the reaction of a diol and a diisocyanate (*Fig 5*).

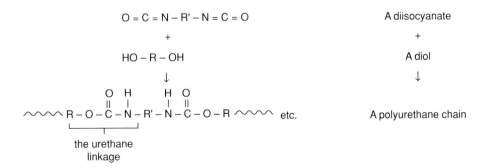

Figure 5 The formation of the polyurethane polymer

Question 2

Can you classify the formation of the polyurethane as an addition or condensation polymerisation? Explain your answer.

RS•C

The chain length of the polymer is controlled by adding some alcohol with just one OH group. The alcohol stops off the ends of the chains.

The R group of the diol contains an alkene functional group (C=C). This group allows the linear polyurethane chains to be crosslinked during curing making the resulting material solidify.

Question 3

Explain how adding alcohols with one OH group stops off the chains. What happens to the average chain length if more of one of these alcohols is added?

How does crosslinking solidify the material?

Crosslinking can be done by reaction with a diperoxide R"–O–O–R'''–O–O–R" which generates free radicals by homolytic fission of the O–O bonds (*Figs 6* and *7*). The reaction is accelerated by a cobalt-based catalyst. This allows the crosslinking reaction to take place in a reasonable time at a temperature below 380 K, above which, damage to cable insulation is possible.

R" – O – O – R''' – O – O – R"　⟶　R" – O˙　˙O – R''' – O – O – R"

R" – O˙　˙O – R''' – O˙　˙O – R"

Figure 6 The peroxide radical generator

RS•C

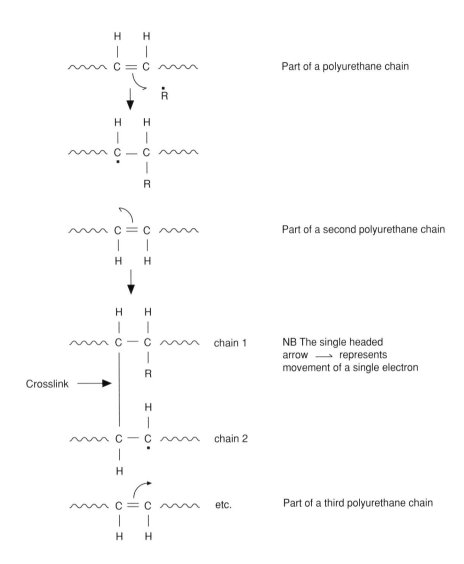

**Figure 7 Crosslinking of the polyurethane molecules
via a free radical mechanism**

Question 4
What is meant by homolytic fission. Why are the O–O bonds expected to break in this way?

The research and development chemists came up with the idea of a two part product. Part A would consist of the polyurethane polymer and solid peroxide in the form of pellets. Part B would contain the cobalt-based catalyst, also in pellet form. The two could be mixed (in a 9:1 ratio) in the solid state and would not react until the two sets of pellets were melted together. This system of pre-mixed solid polymer and catalyst avoids the need for mixing and measuring on site.

In fact the system was slightly more complex than this. It was necessary to delay the crosslinking reaction to prevent it occurring immediately after mixing and allow time for the liquid polyurethane to penetrate between the cables. This was done by adding a small amount of a radical scavenger to part B. This mopped up free radicals generated by the peroxide until all the scavenger was used up. After this the reaction could proceed. The scavenger also acted as an antioxidant thus increasing the shelf life of the product.

Question 5

How could the delay in starting the crosslinking reaction be varied? Suggest a practical reason why it might be useful to be able to vary this lag time.

The scale up process

Once the system described was set up in the laboratory, the challenge became that of transferring the system to high volume production in the factory. As customers were waiting for the product, production had to be maintained during the scale up process. A further consideration was that as ACBS was a fairly small product in the Raychem range, it was desirable to use or adapt existing manufacturing equipment which could still be used for making other products when ACBS was not being manufactured.

The basic manufacturing method is summarised in *Fig 8*.

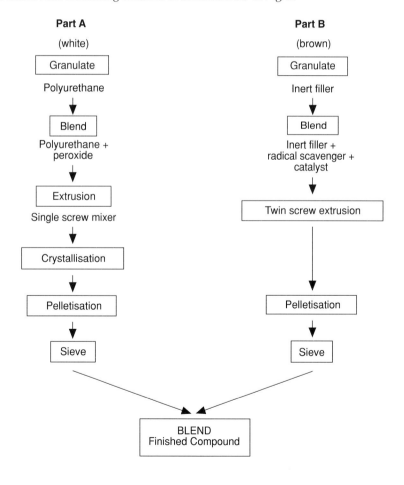

Figure 8 Flow chart for the manufacture of the ACBS cable blocking compound

RS•C

Part A consists of solid polyurethane base and solid peroxide. The polyurethane is obtained from the supplier in the form of lumps about the size of a bar of toilet soap, and is first granulated. The polyurethane and peroxide are then mixed and blended in a blender consisting of a metal screw turning inside a metal cylinder, rather like a mincing machine (*Fig 9*). Heat is generated by the mixing process and this, along with heat from external heaters, melts the two solids. The resulting liquid mixture is forced through an extruder to give a long string of semi-liquid polymer. This is then cooled to leave a long solid rod and cut into cylindrical pellets, about 2 mm in diameter and 2 mm long, by a mechanical pelletiser or cutter. Part B is made in the same way by blending, extruding and pelletising the catalyst and radical scavenger/antioxidant in an inert filler. The filler was chosen so that it did not greatly affect the flow properties of the polyurethane when the two were mixed. White pellets of part A are mixed with brown pellets of part B in a 9:1 ratio and packaged in polythene sachets one face of which is net (with a mesh small enough to retain the pellets). The package is similar to that in *Fig 3*.

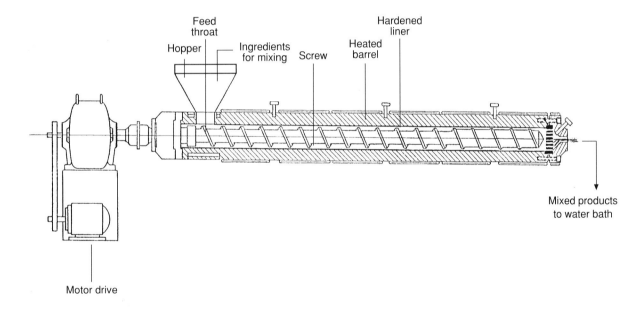

Figure 9 Screw blender

A number of manufacturing problems arose with the process. It was decided to first concentrate on solving those involved with part A as this represented 90% of the material. First, the polyurethane was supplied in large lumps which tended to block the inlet, or throat of the granulator. Secondly, after extrusion, the product was cooled but was slow to solidify, making it impossible to cut the extruded strands into pellets straight after cooling. Normally extruding and cutting are done in a continuous process, the extruded strands cooling in air before being cut by a rotating knife in a pelletising machine. For ACBS, the strands had to be left to cool for 24 hours before they became solid enough to be cut without jamming the pelletising machine. If left longer than this, they would become brittle and break during cutting, giving powder rather than pellets. This would pass through the net in the sachets and be lost, affecting the required 9:1 ratio with the catalyst.

RS•C

This problem slowed down manufacture and required extra plant operators, making the process more expensive. This production problem, of course, was caused by the specification of the material. The long liquid time was vital to the product in use but was also the cause of the manufacturing problem. So, the manufacturing problem could not be resolved by redesigning the material.

Raychem therefore passed the scale up problem to the Chemical Plant Pre-Production group whose objectives were:

▼   to identify a process for manufacturing of the cable block compound;

▼   to make the process capable of large volume production; and

▼   to meet high standards of quality every time the product was produced.

The first manufacturing problem was relatively easily solved. The manufacturers of the polyurethane base material were asked to supply the product in granular form in sizes suitable for the mixer. A specification was developed and the suppliers were able to meet this.

The second problem was more difficult to solve. The first attempt at a solution was to reduce the heating on the part of the mixer closest to the extruder. The extruded strands were then forced into a tank of chilled water at 5 °C (*Fig 10*).

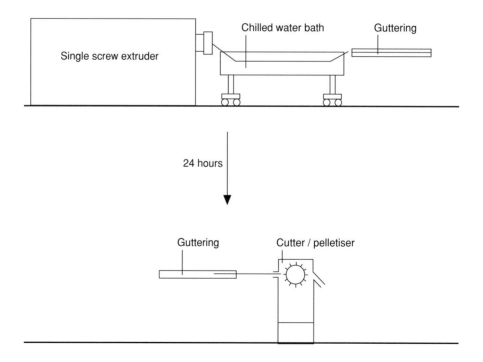

Figure 10 The first attempt at a solution

This resulted in strands of polymer that appeared to be solid but were in fact tubes of solid polymer on the outside filled with an inner core of still-tacky material. This still had to be cut into 2 m lengths and left in guttering for 24 hours to solidify before pelletising. This was, in effect, a batch process rather than a continuous one.

A brainstorming session, led by an external consultant, who had previously worked on ACBS Research and Development, led to the realisation that further cooling was required. Liquid nitrogen was suggested. This was a technique that had been used previously by the R & D department.

RS•C

Liquid nitrogen   **Box 2**

Liquid nitrogen is made by cooling air (by several cycles of compressing it and allowing it to expand and cool) until it liquefies. It is then allowed to heat up until the nitrogen (bpt 77 K) boils off, leaving liquid oxygen (bpt 90 K) behind. The nitrogen vapour is then liquefied again. Liquid nitrogen can be stored in vacuum flasks in which it is continually boiling at its boiling point of 77 K (−196 °C). It is readily available and costs about 10 p per litre.

To try out the idea, a piece of equipment called the Viking helmet (because of its shape) was improvised (*Fig 11*). This was inserted into the production line just before the pelletiser so that the strands were pulled continuously from the extruder, through the water bath, then the Viking helmet and into the cutter. The strands passed through the cold nitrogen vapour above the liquid nitrogen and not through the liquid nitrogen itself. First trials showed an improvement in pelletisation for short periods before the pelletiser clogged. Over-cooling could take place which caused the strands to become brittle and break. Overall, the system showed enough improvement for it to be worth designing a purpose-built liquid nitrogen-cooled vapour bath with valves to control the liquid nitrogen flow. A series of trials with this bath enabled conditions to be found under which batches could be reliably made (*Fig 12*).

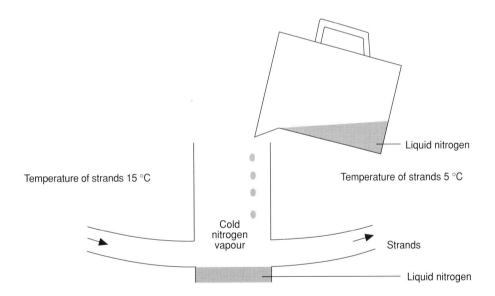

Figure 12 Viking helmet

RS•C

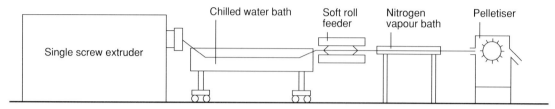

Figure 12 Production line with liquid nitrogen bath

Up to this point, the polyurethane base material and the peroxide had been weighed and blended by hand. The next step was to use weigh feeders in which the two ingredients are continuously fed from bins and the weights which are dispensed are automatically monitored and controlled to maintain the correct ratio of the two materials. Also inserted into the line were two soft roller motorised feeds. These are conveyor belts which help to pull the strands of polymer from the water bath and into the nitrogen cooler and from the cooler into the pelletiser (*Fig 13*).

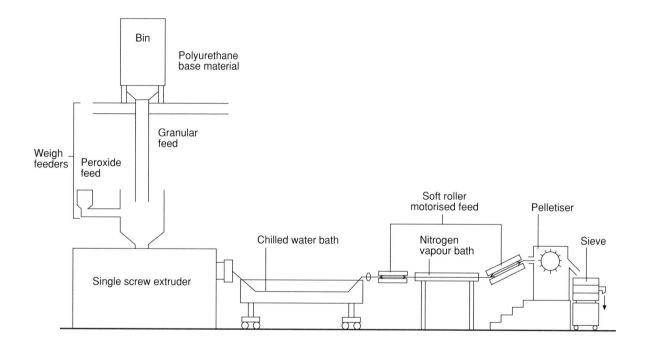

Figure 13 The optimised production line

A timescale for various trial processes is given in *Fig 14*.

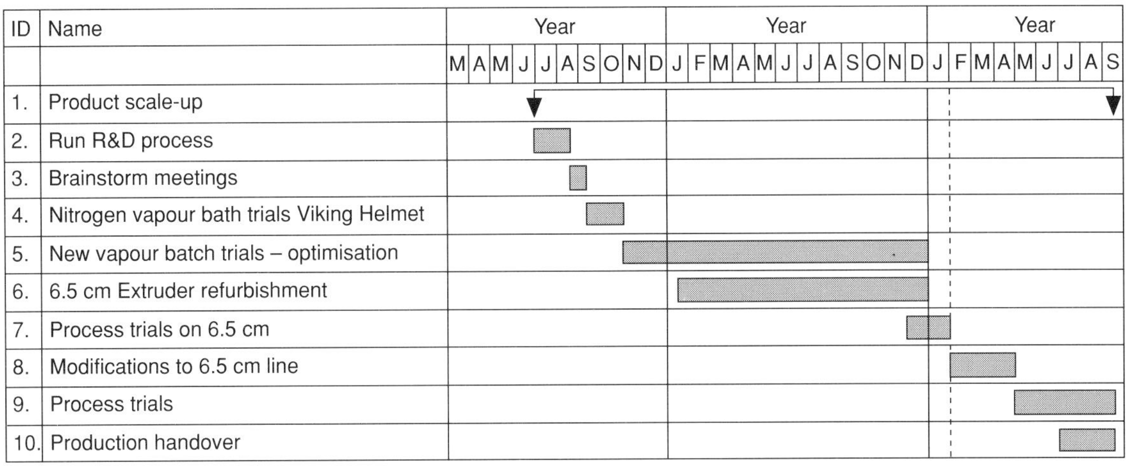

ID	Name	Year										Year												Year								
		M	A	M	J	J	A	S	O	N	D	J	F	M	A	M	J	J	A	S	O	N	D	J	F	M	A	M	J	J	A	S
1.	Product scale-up				▼																											▼
2.	Run R&D process					▓																										
3.	Brainstorm meetings						▓																									
4.	Nitrogen vapour bath trials Viking Helmet						▓	▓																								
5.	New vapour batch trials – optimisation							▓	▓	▓	▓	▓	▓	▓	▓	▓	▓	▓														
6.	6.5 cm Extruder refurbishment											▓	▓	▓	▓	▓	▓	▓	▓													
7.	Process trials on 6.5 cm																			▓	▓											
8.	Modifications to 6.5 cm line																							▓	▓							
9.	Process trials																									▓	▓					
10.	Production handover																												▓	▓		

Figure 14 Timescale for scale-up

Once the Part A process had reached this stage, the Chemical Pre-Production group handed it over to the Production Group but continued to provide support and staff training. They then moved on to develop the process for making part B which had a number of similar problems.

During this stage of process development, the cost of manufacture of part A dropped by about 50% and later improvements reduced this by a further 50% of the new cost.

Postscript

Since the stage described above, further improvements have been made. The nitrogen cooling system was expensive to operate as it used large quantities of liquid nitrogen and the overall process was very sensitive to small changes in operating conditions. This meant that it was difficult to run with normal production personnel. Eventually new types of pelletiser became available which had been developed to pelletise adhesives which were soft. These proved capable of pelletising ACBS part A without nitrogen cooling. A system was developed whereby the mixture was cooled while still in the extruder and was extruded in an almost solid state into chilled water. This system is more flexible since it can be used for several different product applications whereas the nitrogen cooling system was specific to this one product. However, the knowledge of the nitrogen based system is still available should it be required for a product in the future.

RS•C

Answers to questions

1.

The monomers have a functional group on either end.

2. Addition, as no small molecule is eliminated.

3. To form a polymer, both monomers must have a functional group on each end of the molecule. An alcohol with just one OH group can react at one end only. The unreactive end of the molecule then forms the end of the polymer chain. The addition of more monofunctional alcohol reduces the average chain length and adding less increases it.

Crosslinking prevents the chains sliding over one another and thus makes the resulting structure more solid.

4. Bond breaking in which the two electrons in a single bond go on to each fragment thus forming a pair of free radicals. This contrasts with heterolytic fission where both electrons go to one fragment and none to the other thus forming a positive and a negative ion.

Homolytic bond fission is expected in this case as the two atoms which form the bond are the same (oxygen) and therefore have the same electronegativity and thus equal attraction for the bonding electrons. The O–O bond is likely to break as it is relatively weak, 144 kJ mol^{-1}, compared with a typical covalent bond such as C–C at 347 kJ mol^{-1}.

5. Adding more scavenger increases the lag time. This gives longer for the blocking agent to penetrate between the pairs of cable. This might be useful in cold conditions where the blocking agent is more viscous or in situations with many pairs of cables tightly packed together. It might also be useful for blocking a large cable where a longer liquid time is required for the blocking agent to penetrate all the cable pairs.

RS•C

Radioactive discharges into the River Ribble

British Nuclear Fuels (BNFL) Fuels Division at Springfields near Preston makes nuclear fuel for different types of nuclear reactors – the Magnox, advanced gas cooled reactors (AGR) and the pressurised water reactors (PWR). The process begins with uranium ore concentrate (UOC) imported from around the world. This is made from uranium ore obtained by open cast mining. The original ore contains on average about 1.5% uranium (although this is quite variable and can be as low as 0.15%) and is ground, treated and purified until the UOC contains about 80% uranium.

During the processing at Springfields, some waste from the plant is discharged into the nearby River Ribble. The company has recently completed a survey into ways of minimising discharges of radioisotopes.

The processes for making nuclear fuel

Magnox power stations use fuel elements which consist of natural uranium rods encased in a magnox can (magnox is an alloy of 99% magnesium and 1% aluminium). The fuel for both AGRs and PWRs is enriched uranium oxide (UO_2) in the form of pellets stacked inside either stainless steel (AGR) or zirconium alloy (PWR) fuel pins (*Fig 1*). A group of fuel pins is arranged together to form an assembly.

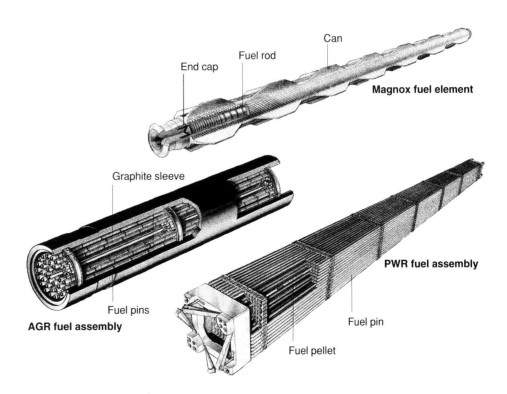

Figure 1 Fuel rods and pins

RS•C

Natural and enriched uranium   **Box 1**

Natural uranium consists almost entirely of two isotopes, 99.3% ^{238}U and 0.7% ^{235}U. ^{235}Uranium readily undergoes fission when hit by a neutron – giving out energy, two smaller elements and two further neutrons which can initiate further fission and bring about a chain reaction. One possibility is

$$^{235}U + {}^{1}n \rightarrow {}^{144}Ba + {}^{90}Kr + 2{}^{1}n$$

^{238}Uranium does not undergo fission in this way. Since there is a high percentage of non-fissile ^{238}U in magnox fuels, it is important that few neutrons are lost and the fuel element casings must be poor neutron absorbers or the ^{235}U chain reaction might stop. This is the reason for the choice of the particular alloy.

Magnox reactors use natural uranium while AGRs and PWRs use enriched uranium-based fuels – to 2.3% ^{235}U in an AGR and to 3.2% in a PWR. Enrichment is done in a centrifuge. Gaseous uranium hexafluoride (UF_6) is spun in a centrifuge and the heavier $^{238}UF_6$ molecules are flung outwards more than the lighter $^{235}UF_6$ ones.

On arrival at Springfields the UOC, containing U_3O_8 and other uranium compounds, is first dissolved in concentrated nitric acid to give a solution containing uranyl nitrate ($UO_2(NO_3)_2$) and other nitrates.

$$U_3O_8(s) + 8HNO_3(aq) \rightarrow 4H_2O(l) + 3UO_2(NO_3)_2(aq) + 2NO_2(g)$$

Insoluble impurities, such as sand, can be filtered off. The solution is then solvent extracted with a solution of tributyl phosphate in kerosene. Uranyl nitrate is soluble in this solution but the other nitrates are not. Further extractions produce a solution of 99.95% purity. After extraction into the solvent, the uranyl nitrate is back extracted using slightly acidified water and the solvent is re-used.

The slightly acidified solution of uranyl nitrate is then evaporated and the uranyl nitrate is thermally decomposed to produce uranium trioxide (UO_3).

$$2UO_2(NO_3)_2(s) \rightarrow 2UO_3(s) + 4NO_2(g) + O_2(g)$$

Question 1

What problem would be caused by the nitrogen dioxide given off in two of the processes above if it were discharged into the atmosphere? Suggest a useful product which might be made from this gas.

Uranium trioxide is then reduced with hydrogen to give uranium dioxide (UO_2)

$$UO_3(s) + H_2(g) \rightarrow UO_2(s) + H_2O(l)$$

and then reacted with hydrofluoric acid to produce uranium tetrafluoride (UF_4), a green powder.

$$UO_2(s) + 4HF(l) \rightarrow UF_4(s) + 2H_2O(l)$$

RS•C

This is the starting material for two processes, one of which produces natural uranium metal for magnox fuel and the other which produces enriched uranium dioxide for AGR and PWR fuel.

To make uranium for magnox fuel, the uranium tetrafluoride is mixed with magnesium chippings and heated electrically in a graphite container under an argon atmosphere (which prevents oxidation of the uranium as it is formed). Molten uranium is produced along with a slag of magnesium fluoride which floats on top. This slag is separated and the molten uranium cast into rods.

$$UF_4(s) + 2Mg(s) \rightarrow 2MgF_2(s) + U(l)$$

To make enriched uranium dioxide fuel, the uranium tetrafluoride is first heated with fluorine gas to produce uranium hexafluoride (hex), boiling point 60 °C. UF_4 is ionic, U^{4+} $4F^-$, while UF_6 is molecular. This is then enriched as described in Box 1. Enrichment is carried out at Capenhurst near Chester, the hex being transported there and back by road tanker. The hex is then converted to uranium dioxide by reaction with steam followed by reaction with hydrogen, all in a single stage. Uranium dioxide powder is then pressed into fuel pellets. The reactions are

$$UF_4(s) + F_2(g) \rightarrow UF_6(g)$$

$$UF_6(g) + 2H_2O(l) + H_2(g) \rightarrow UO_2(s) + 6HF(l)$$

Question 2

Work out the oxidation number of the uranium on both sides of the above equation. What has happened to the uranium in the reaction? What is the function of the hydrogen in the reaction?

The environmental problem

Waste from the Springfields site is discharged into the tidal estuary of the River Ribble via a pipeline (*Figs 2 and 3*).

RS•C

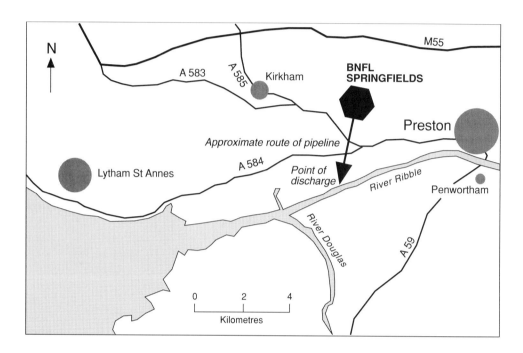

Figure 2 The location of the Springfields site and the discharge pipeline

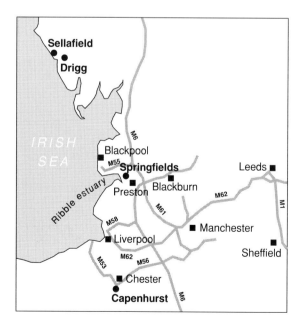

Figure 3 The locations of Springfields, Sellafield, Drigg and Capenhurst

This waste includes both conventional chemical waste and a number of radioactive isotopes – principally of uranium and thorium – which vary in type of radioactivity (α or β), activity and half life. The amount of radioactive discharge permitted is

RS•C

governed by a certificate of authorisation issued jointly by Her Majesty's Inspectorate of Pollution (HMIP) and the Ministry of Agriculture, Fisheries and Food (MAFF) under the Radioactive Substances Act, 1960.

Note. HMIP is now part of the Environment Agency (EA) and any new Certificates of authorisation will be issued by the EA with MAFF as consultees.

The principal isotopes involved at Springfields are shown in Table 1. They derive from uranium isotopes in the ore and thorium impurities and their daughter products produced by radioactive decay.

Isotope	Type of radiation	Half life
^{230}Th *	α	8×10^4 years
^{232}Th	α	1.4×10^{10} years
^{238}U	α	4.5×10^9 years
^{234}Th *	β	24.1 days
^{234m}Pa *	β	1.17 min

* ^{234}Th, ^{234m}Pa and ^{230}Th are daughter products of the decay of ^{238}U. ^{232}Th is a radioactive impurity present in the ore on arrival at Springfields. ^{234m}Pa is an excited state of the ^{234}Pa isotope.

Table 1 The principal isotopes involved at Springfields

Question 3

Write a nuclear equation for

a)   the α decay of ^{230}Th

b)   the β decay of ^{234}Th.

What new element is formed in each case?

It is important to realise that there is no way of chemically changing one radioisotope into another (possibly less intensely radioactive or with a shorter half life). The only options are to chemically remove radioisotopes from the waste for storage (short term in the case of short-lived isotopes, long term for those with longer half lives) or dispersal of the radioisotopes.

The Ribble estuary is tidal and also receives discharges from Sellafield further up the coast to the north (*Fig 3*). The people considered by MAFF to be most at risk from radioactive discharges in the estuary – the critical group – are ones who live in houseboats moored in the estuary and also anglers and wildfowlers.

In 1991 BNFL Springfields was required by MAFF and HMIP to carry out a survey of their radioactive discharges to develop a Best Practicable Environmental Option (BPEO) to minimise pollution and protect the environment. This was completed in 1993. It involved a study of the processes on the Springfields site to identify which pollutants (both chemical and radioactive) were generated where, and measurements

RS•C

of their amounts. A number of options was then identified for reducing discharges. Each option was then examined. This was done in two stages. First, each was assessed against legal, technical and financial constraints to reject any non-feasible options. The remaining options were then scored against a number of criteria, each criterion being weighted according to its importance. The criteria used to identify the final chosen method included cost, timescale, environmental impact and technical feasibility.

Identifying the sources of the pollutants

The Springfields plant was divided into 13 areas. These varied from production process areas such as the plant where the uranium ore concentrate is dissolved in nitric acid to the laundry where contaminated lab coats and overalls are washed. These produced a total of 61 effluent streams (not literally open streams). Analysis of these streams showed that 99.8% of the thorium came from just one stream – waste from the solvent extraction of uranyl nitrate. 29.1% of uranium came from this source but there were also significant amounts from other processes. These included other uranium processing plants, the laundry, the decontamination centre and spillages from waste storage areas. Most of the effluent is significantly acidic.

Options for treatment

Twenty-one detailed options were considered including a no change option (option 21). They are listed in Table 2. They fell into four main groups.

1.   To treat the effluent to remove radioisotopes as solid waste for disposal off-site (options 1-14 in Table 2).

2.   To store isotopes of short half-life on site until their radioactivity had decayed to a less-problematic level (options 15-17 in Table 2).

3.   To limit the impurities taken onto the site, *ie* change the specification for the UOC raw material so that less thorium is allowed (option 18 in Table 2).

4.   To continue to discharge radioisotopes but in a different way so that they became better dispersed in the estuary or carried out to sea more effectively (options 19 and 20 in Table 2).

RS•C

No.	Description	Effluent treated	Output	Status
1	Direct fluorination	None – new method of making UF_6	Dry solids	lab
2	Solvent extraction	All Th isotopes	ILW, waste solvent	pp
3	Ion exchange	All U & Th isotopes	LLW, ion exchange material, waste liquor	lab
4	Reverse osmosis	All U & Th isotopes	Moist solids, waste liquor	lab
5	Solar evaporation	All U & Th isotopes & other solids	Moist alkaline solids	we
6	Precipitation & clarification	All U & Th isotopes & other solids	Moist alkaline solids & alkaline liquor	we
7	Sulfide/oxalate/ peroxide precipitation	All U & Th isotopes & heavy metals	Moist solid salts	lab
8	Fluoride precipitation	Th isotopes	LLW & neutral liquor	lab
9	Addition of lime & solids removal	All U & Th isotopes & heavy metals	LLW & alkaline liquor	we
10	Addition of NaOH	All U & Th isotopes & heavy metals	LLW	we
11	Evaporation & disposal	All U & Th isotopes & nitrates	Moist solids, HNO_3 for re-use, water	we
12	Evaporation & neutralisation	All U & Th isotopes, heavy metals, nitrates & suspended solids	LLW, HNO_3 for re-use, water	we
13	Evaporation & calcination	All U & Th isotopes, heavy metals, nitrates & suspended solids	Dry solids, HNO_3 for re-use, water	we
14	Vitrification	All U & Th isotopes, heavy metals, nitrates & suspended solids	LLW as glass blocks, HNO_3 for re-use, water	we
15	β decay storage	Th β activity	Normal with less β activity	we
16	Evaporation & β decay storage	Nitrates & Th β activity	Alkaline slurry with less β activity	we
17	Neutralisation & β decay storage	Add lime and store the resulting slurry	Alkaline slurry with less β activity	we
18	Selective UOC processing	Th activity	Normal but with less Th content	we
19	Discharge timed to tide	All	Alkaline slurry	we
20	Extend pipeline 6 km downstream	All	Alkaline slurry	we
21	No change	–	–	–

Key

lab	laboratory stage
pp	pilot plant
we	well established
ILW	intermediate level radioactive waste
LLW	low level radioactive waste

Table 2 The options for treatment

RS•C

Question 4

Why does storage (options 15–17) significantly reduce thorium β activity but not thorium α activity?

It is worth noting that, since the object of the company is the production and sale of uranium, any recovered uranium is potentially valuable. This is not so for thorium as there is only a small market for this element and its compounds (in the manufacture of gas mantles).

Most of the options considered fall into group 1. They included removal of the radioisotopes from the waste streams by:

▼   solvent extraction (Box 2);

▼   reverse osmosis (Box 3);

▼   ion exchange (Box 4);

▼   evaporation, followed by vitrification – heating with glass to form a solid glass matrix; and

▼   precipitation in a variety of ways, for example as insoluble metal fluorides, metal hydroxides, sulfides, oxalates (ethanedioates) or peroxides.

RS•C

Solvent extraction   Box 2

This method of separation is carried out in the laboratory by shaking a solution of a solute in one solvent with another solvent which is immiscible with the first. The two solvents are then allowed to separate. The solute will distribute (or partition) itself between the two solvents according to its solubility in the solvents. For example, iodine is about 80 times as soluble in trichloroethane as in water so if an aqueous solution of iodine is shaken with trichloroethane and the layers are allowed to separate, the concentration of iodine in the trichloroethane layer is 80 times that in the aqueous layer. The partition coefficient of iodine between the two solvents is given by

$$\frac{[I_2(CH_3CCl_3)]}{[I_2(aq)]} = 80$$

Industrially, the extraction is done by a continuous flow process in which the two solvents (tributyl phosphate (TBP) in kerosene, and water) flow in opposite directions through a series of mixer-settler boxes (see below).

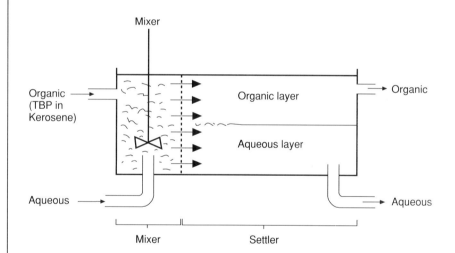

A mixer settler chamber

The chambers are arranged "back to back" so that the aqueous and the organic layers flow in opposite directions and uranyl nitrate is gradually transferred from the aqueous to the organic layer.

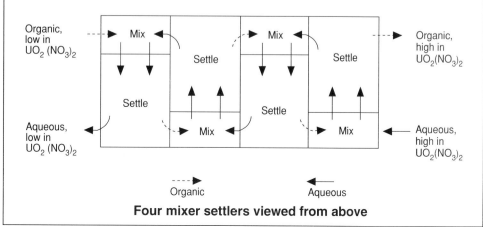

Four mixer settlers viewed from above

RS•C

Each of these methods produces a solid waste incorporating the radioisotopes of uranium and thorium. Most of the products are low level waste (LLW) which can then be taken for disposal to the radioactive waste facility at Drigg, Cumbria. However, some produce intermediate level waste (ILW) for which there is no existing commercial disposal facility in this country and BNFL would have had to store such waste themselves. (Nirex is currently seeking a suitable ILW storage site). One issue which was considered when selecting the options was that Drigg's capacity is finite and that it is preferable to find methods which did not use this national resource.

Reverse osmosis   **Box 3**

This is a process of separating a solute from a solution by using pressure to force the solvent through a semi-permeable membrane or SPM (one which allows small solvent particles to pass through but not the larger solute particles). This leaves a concentrated solution of the solutes (including, in this case, ammonia, nitrates and the radioisotopes) and the almost-pure solvent. This is the reverse of the process which occurs in osmosis where the SPM separates a concentrated solution from pure solvent and the solvent diffuses through the SPM in an attempt to equalise the concentrations.

Ion exchange   **Box 4**

Ion exchange involves feeding the solution containing the radioisotopes into a column containing a suitable ion exchange material. This contains ions such as sodium ions which the material exchanges for ions of the radioisotope. A similar process is used to soften hard water but in this case calcium ions (which cause hardness) are exchanged for sodium ions (which do not). The exchanged ions of the radioisotope can be freed from the resin and the resin regenerated for re-use by washing the resin with a concentrated solution of sodium sulfate which reverses the exchange process. The ions of the radioisotope can then be precipitated from the sulfate solution to give a solid waste for disposal.

Selecting the option

Each option was listed against the effluent streams it was suitable for treating, the output of the process (*eg* low level waste) , its development status (*eg* already in use, laboratory stage *etc*) and the level of expertise available within BNFL (*Fig 4*).

RS•C

Option Reference Number	3
Effluent Treatment Option	Ion Exchange
Effluent Treated	Radionuclides (U and/or Th – all isotopes)
Input Stream Numbers	1 & 2
Description of Effluent Treatment	The liquor is fed into a column containing a suitable ion exchange material which extracts thorium preferentially to give a de-thorinated residue which is neutralised. The loaded ion exchange material is then backwashed using a sodium sulfate solution to regenerate the ion exchange material for re-use. The thorium is then precipitated from the sodium sulfate solution to produce a solid waste for disposal.
Outputs	Approximately 320 tonnes per year of encapsulated LLW, waste ion exchange material, neutralised raffinate liquor.
Development Status	Operated at laboratory scale for Th isotopes only within BNFL Fuel Division.
Expertise	Limited exerience at laboratory scale for Th isotopes within BNFL Fuel Division.

Figure 4 A typical option specification

The coarse screen

Each option was first assessed against the technical, legal and financial constraints involved in its use. This was seen as a coarse screen to reject any non-feasible options. Remaining options were then scored against a list of weighted criteria (*Fig 5*)

RS•C

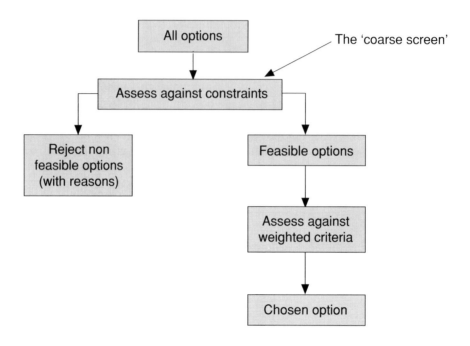

Figure 5 Coarse screening

Constraints

Legal

Discharges of liquid radioactive waste from the Springfields site are governed by a Certificate of Authorisation specifying, among other things, the quarterly and annual amounts shown in Table 3.

Type of activity	Quarterly limit (TBq)	Annual limit (TBq)
α activity	1.5	4
β activity	80	240
^{230}Th	Not specified	2
^{232}Th	Not specified	0.2
Uranium	Not specified	0.15
^{237}Np	Not specified	0.04
^{99}Tc	Not specified	0.60

Table 3 Radioactive discharge limits for the Springfields site
Note: TBq stands for tera becquerels (the prefix tera means 10^{12}) where one becquerel represents one nuclear disintegration per second

RS•C

Another set of conditions governs the disposal of solid waste at the Drigg site. This defines the total activity, the activity of different radioisotopes, the total volume and the chemical state of the isotopes.

No options were rejected on legal grounds.

Technical

These were largely based around the inability to develop the technology within the required time scale or, in two cases involving solar evaporation, unsuitability of climate.

Six options were rejected on technical grounds – nos. 1, 4, 5, 6, 7 & 11.

Financial

It was decided that any expenditure incurred should be proportional to the benefit to be derived from [it], so a cost-benefit analysis was undertaken. This was done on the basis of work done by the National Radiological Protection Board (NRPB) which has derived a monetary value for the detriment to an individual of exposure to radiation. In other words, the NRPB has tried to estimate the economic costs of exposure to different levels of radiation. These include medical costs for anyone made ill by exposure and loss of production caused by people being off work. The result of this was a recommendation that £20 000 per year should be the maximum amount of money to be spent. This low figure is a reflection of the relatively low levels of discharge from Springfields.

For each option, an annual cost was estimated by making the following assumptions.

▼ Any capital outlay, such as a treatment plant, would have a lifetime of 24 years, 20 of which are working (the others would be building and decommissioning).

▼ Decommissioning costs (*ie* demolishing and making safe the plant after its useful life is over) would amount to 25% of initial capital cost.

The costs of the remaining options are shown in Table 4.

RS•C

Option No.	Option	Cost £000 per year
2	Solvent extraction	3460
3	Ion exchange	3990
8	Fluoride precipitation	2950
9	Addition of lime with solids removal	2390
10	Addition of caustic soda with solids removal	9080
12	Evaporation and neutralisation	6040
13	Evaporation and calcination	3310
14	Vitrification	4850
15	β decay storage of liquor	2670
16	Evaporation and decay	1340
17	Neutralisation and decay storage	940
18	Selective UOC processing	40
19	Tidal discharge	40
20	Extended pipeline downstream	70
21	Current discharges to the River Ribble*	0*

* Note that the actual cost of operating the current discharge system is in fact £440 000 per year but this cost has to be incurred for other reasons and it is not an extra cost as the others would be.

Table 4 Costs of the remaining options

All options, except the no change option, are estimated to cost more than £20 000 per year. Even so, the four cheapest were taken on to the final scoring process.

Scoring the remaining options

The four cheapest options were:

▼   No. 18, selective ore processing;

▼   No. 19, tidal discharge;

RS•C

▼   No. 20, extending the pipeline downstream; and

▼   No. 21, no change.

These were scored against each of the criteria below. Each criterion was then given a weighting (shown in brackets) which indicated its level of importance. The score was then multiplied by its weighting to give a weighted score. The weighted scores for all the criteria were then added up to give a total weighted score.

Criteria

▼   Dose uptake to the human population (0.40) (This was actually broken down into three subcategories – collective dose (0.04), surface dose (0.02) and committed effective dose equivalent, CEDE, (0.34). See Box 5 for an explanation of the different doses.)

▼   Lifetime cost (0.25).

▼   Environmental effect on flora and fauna (0.15).

▼   Time scale for implementation (0.10).

▼   Wider social issues (0.10).

Note that the total of the weightings = 1.00.

Dose uptakes to human population   **Box 5**

Collective dose refers to the population of the world. It is measured in man S sieverts per year (man Sv/a)

Critical group dose is the dose to the most exposed group of people (the critical group) *ie* boat dwellers, wildfowlers and fishermen. It is measured in μSv.

Committed effective dose equivalent (CEDE) is calculated from three factors:

▼   dose from ingestion or inhalation of radioactive material;

▼   whole body dose from penetrating (γ) radiation; and

▼   dose to particular parts of the body such as the gonads in males.

CEDE is also measured in μSv.

The sievert, Sv, is a unit of dose equivalent which takes into account the fact that different types of radiation (α, β, γ) have different biological effects. 1 Sv is equivalent to an energy of 1 J/kg.

Devising weightings was problematic. They were produced by canvassing a wide spectrum of people across the company. This gave an industrial bias to the weightings but it was felt that the overall order of importance was correct.

The scoring procedure was also difficult to devise. Each criterion was scored out of 100 such that the best performing option scored 100 and the worst performing 0. The scoring was done graphically. For example, the cost of the four remaining options varied between £0 and £70 000. These were placed on a graph so that cost £0 was given a score of 100 and cost £70 000 a score of 0. So an option costing £40 000 would get a score of 43 (*Fig 6*).

RS•C

Cost
Weighting   0.25
Range of Values   0–70,000 £ per year
Graph used

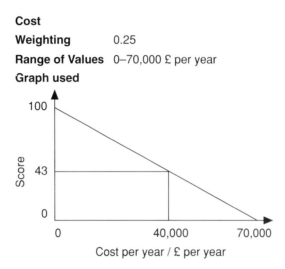

Figure 6 Scoring the costs of different options

Other criteria were even harder to quantify so that the horizontal axis on the wider social concerns graph (*Fig 7*) could only be labelled LOW – HIGH.

Wider Social Concerns
Weighting   0.10
Range of Values   indicates increasing concern
Graph used

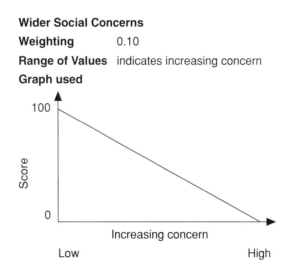

Figure 7 Scoring social concern

However, the scoring system was the same, the best option scoring 100 and the worst 0.

Table 5 shows the actual data for each of the four remaining options and Table 6 shows these after being translated into scores out of 100 from the graphs.

RS•C

Option No.	Criteria						
	Collective dose manSv per year	Surface dose µSv	CEDE µSv	Time-scale yrs	Cost £000 per year	Env impact	Wider social concerns
18	0.09	12,000	93	2.0	40	Low	Med/high
19	0.16	8,700	440	2.5	40	Med	Low
20	0.16	6,300	330	2.5	70	Med/high	Med/low
21	0.16	12,000	430	0	0	High	High

Table 5 Criteria values for each of the final four options

Option No.	Criteria						
	Collective dose (weighting 0.04)	Surface dose (weighting 0.02)	CEDE (weighting 0.34)	Time-scale (weighting 0.10)	Cost (weighting 0.25)	Env impact (weighting 0.15)	Wider social concerns (weighting 0.10)
18	100	0	100	20	43	100	25
19	0	58	0	0	43	50	100
20	0	100	32	0	0	25	75
21	0	0	3	100	100	0	0

Table 6 Scores for each of the remaining options

Tables 7–10 show the actual scores for the four remaining options after applying the weightings.

Criterion	Score	Weighting	Weighted score
Collective dose	100	0.04	4
Surface dose	0	0.02	0
CEDE	100	0.34	34
Timescale	20	0.10	2
Cost	43	0.25	10.75
Environmental impact	100	0.15	15
Wider social concerns	25	0.10	2.5
Total weighted score			**68.25**

Table 7 Option 18: selective UOC processing

Criterion	Score	Weighting	Weighted score
Collective dose	0	0.04	0
Surface dose	58	0.02	1.16
CEDE	0	0.34	0
Timescale	0	0.10	0
Cost	43	0.25	10.75
Environmental impact	50	0.15	7.5
Wider social concerns	100	0.10	10
Total weighted score			**29.41**

Table 8 Option 19: tidal discharge

RS•C

Criterion	Score	Weighting	Weighted score
Collective dose	0	0.04	0
Surface dose	100	0.02	2
CEDE	32	0.34	10.88
Timescale	0	0.10	0
Cost	0	0.25	0
Environmental impact	25	0.15	3.75
Wider social concerns	75	0.10	7.5
Total weighted score			24.13

Table 9 Option 20: extend pipeline downstream

Criterion	Score	Weighting	Weighted score
Collective dose	0	0.04	0
Surface dose	0	0.02	0
CEDE	3	0.34	1.02
Timescale	100	0.10	10
Cost	100	0.25	25
Environmental impact	0	0.15	0
Wider social concerns	0	0.10	0
Total weighted score			36.02

**Table 10 Option 21: continuation of discharges to the Ribble
– the no change option**

These show that selective UOC processing (with a score of 68.25/100) is the preferred option. The next highest score is the no change option with 36.02/100.
    Selective UOC processing involves re-negotiating contracts with the suppliers to reduce the limit for thorium in the UOC from 2.5% to 0.25%. This automatically

RS•C

reduces dischages of ^{232}Th and to some extent ^{230}Th but does not affect the levels of ^{234}Th and ^{234}Pa as these are relatively short-lived daughter products of uranium and are actually formed on site.

In fact selective UOC processing was partially implemented during the Best Practicable Environmental Option (BPEO) study, after 1991, and the results are shown in *Fig 8*. These show roughly a six-fold reduction in the discharges of ^{230}Th and a four-fold drop in ^{232}Th levels.

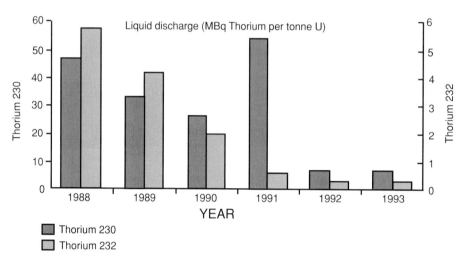

Note: The becquerel (Bq) is a unit of radioactive decay of one decay per second

Figure 8 Effect of selective UOC processing on discharges to the Ribble

---

Mathematical modelling   **Box 6**

To better understand the effects of the options which involve changing the method of discharge, a mathematical model of the estuary was commissioned. As yet, this has failed to correctly predict known observations. Further work is required.

Bibliography

The following publications give more information on the manufacture of nuclear fuels and the other activities at the Springfields site.

Springfields, Risley, Warrington: BNFL, 1992.

Fuel Manufacturing Technology, Risley, Warrington: BNFL, 1991.

Manufacturing Nuclear Fuels (Hobsons Science support Series), Hobsons: Cambridge, 1988.

Answers to questions

1. Nitrogen dioxide could react with water and oxygen in the air to form nitric acid – a cause of acid rain. The NO_2 fumes are passed to a nitric acid recovery plant on site and the nitric acid is re-used in the Springfields complex.

2. U(VI) on the left, U(IV) on the right. The uranium has been reduced. The hydrogen is a reducing agent (being oxidised from 0 to +I in the process).

3. a) $^{230}_{90}Th \rightarrow {}^{4}_{2}He + {}^{226}_{88}Ra$

   The new element formed is radium.

   b) $^{234}_{90}Th \rightarrow {}^{0}_{-1}e + {}^{234}_{91}Pa$

   The new element is protactinium.

4. The thorium α-emitters have such long half lives that storage for the forseeable future will not significantly reduce their activity. The thorium β-emitter has a short half life of only 24 days, so (for example) storage for ten weeks (70 days, approximately three half lives will reduce its activity to almost 1/8 of its original value.

RS•C

The Shell ethylene pipeline

Introduction

In 1992, Shell Chemicals Europe completed the 411 km North Western Ethylene Pipeline (NWEP) between Grangemouth on The Firth of Forth in Scotland and Stanlow in Cheshire (*Fig 1*). The purpose of the pipeline is to transport ethylene (ethene) * obtained from the North Sea oil and gas fields to Shell's petrochemicals manufacturing sites at Stanlow and Carrington, both in Cheshire. The ethylene (boiling point -104 °C) is pumped under pressure in a so-called dense gas phase so that it is almost liquid. This study looks at the issues – chemical, technological, environmental and financial – which had to be considered during the planning, design and building of the pipeline.

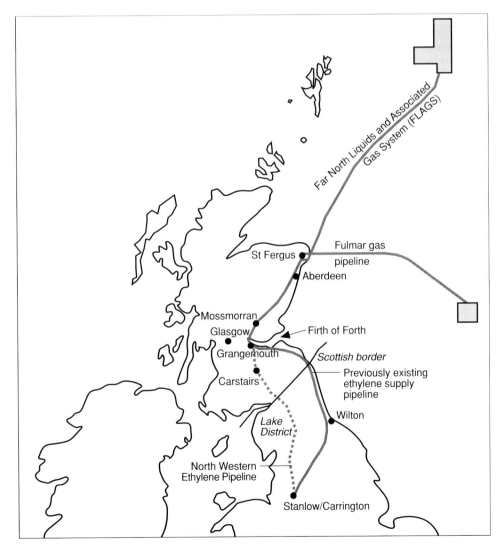

* Note. The systematic name for ethylene ($H_2C=CH_2$) is ethene but it is usually still called ethylene in the petrochemicals industry. The name ethylene will be used throughout this study.

Figure 1 The pipeline's geography

RS•C

Context

Ethylene is an important feedstock for the chemical industry being used to manufacture a wide variety of chemicals (*Fig 2*). Ethylene's importance derives from its chemical reactivity. This is brought about by its double bond which enables ethylene to undergo many reactions, including polymerisation.

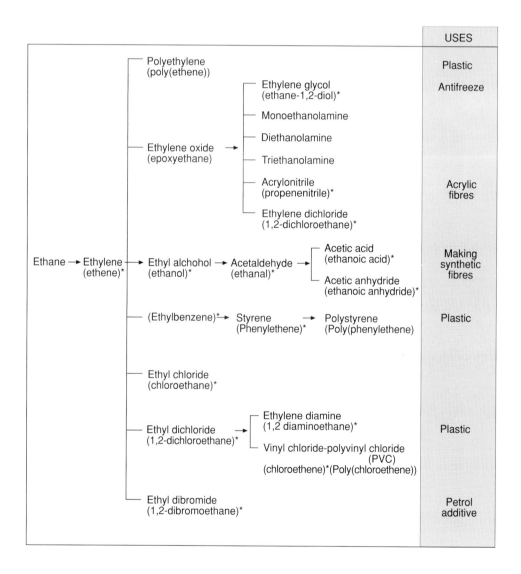

Figure 2 The uses of and products from ethylene

Question 1

What are the structural formulae for the organic compounds marked * in *Fig 2*.

Ethylene is manufactured by cracking, from a variety of hydrocarbon feedstocks which are obtained from North Sea oil and gas. These raw materials come by undersea pipelines from the Brent and Central Gas fields in the North Sea to a terminal at St Fergus, North of Aberdeen. Here, methane is removed and sold to British Gas while the remaining natural gas liquids (NGL), largely short chain alkanes, are transferred south by pipeline to Mossmorran, where they are fractionated into ethane, propane, butane and natural gasoline. Much of the ethane is cracked to ethylene at the nearby cracking plant at Grangemouth (*Fig 3*).

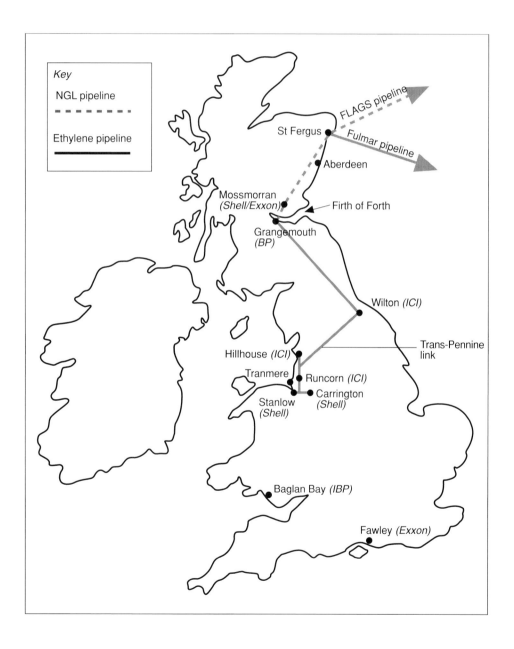

Figure 3 The ethylene system prior to the NWEP pipeline

RS•C

Question 2
Write equations to show the cracking of (a) pentane and (b) ethane to give ethylene (ethene).

Question 3
Write equations for the complete combustion of methane and of ethane. Why are methane and ethane are not interchangeable as domestic fuels?

Suggest a further, economic, reason why ethane is not burnt as a fuel but methane is.

Defining the problem

Economically, the output of the Stanlow and Carrington plants is important to Shell itself and to the local and national economies. Stanlow, for example, contributes around £8 million in rates to the Local Authority. It has been estimated that 75% of chemical producers in the Ellesmere Port area near Stanlow buy feedstock from Stanlow. Around three quarters of the output of the Stanlow plant is exported, contributing to the overall balance of payments surplus of chemical manufacturing as a whole.

Prior to the completion of the pipeline, ethylene was supplied to Stanlow via an existing pipeline running down the East of the UK to Wilton in Cleveland and thence across the Pennines. This pipeline is jointly owned by BP and ICI – commercial competitors of Shell's. Shell had to negotiate (and pay for) the right to transport its ethylene via this pipeline. It was clearly unsatisfactory to Shell to have the supply of a major raw material in the control of other companies. Furthermore, the maximum capacity of this pipeline was limited, and forecasts of demand for ethylene-based products indicated a need for further capacity. Shell predicted an annual requirement of 610,000 tonnes per year while BP/ICI could make available only 450,000 tonnes per year through the existing pipeline after meeting their own requirements.

At Stanlow, the Shell Higher Olefin Plant (SHOP) manufactures long, straight chain olefins (alkenes) from ethylene. These are used as lubricants and in the manufacture of detergents, among other things.

The four plants at Carrington produce:

▼   low density polythene for packaging film, mouldings and coatings;

▼   polypropylene for plastic chairs and crates;

▼   expanded polystyrene for packaging and insulation; and

▼   a variety of other products including detergents and lubricants.

Suggested solutions

Shell considered the following five ideas to solve the ethylene raw material transfer problem.

1.   A no-change scenario with no increase in production at Stanlow and Carrington.

2.   Developing chemical plant to process the ethylene at Mossmorran.

3.   Developing alternative sources of ethylene, other than Mossmorran, to supply Stanlow and Carrington.

4.   Delivering ethylene from Mossmorran to Stanlow by road, rail or sea.

5.   Building a new pipeline

Each alternative was carefully considered.

Option 1

This was rejected on the grounds that it would lead to Shell's having to transfer production elsewhere in the world to meet demand. This would involve a loss of export earnings and an adverse effect on the UK's balance of payments. It was considered that ultimately the Carrington site would have to be shut down causing loss of jobs both at Carrington itself and in other local industries which use materials made at Carrington.

Option 2

Option 2 was rejected (for the present time) as Shell did not wish to build a third UK site with the consequent increase in overheads and fragmentation of their operations, particularly as the Stanlow and Carrington sites already have facilities such as storage tanks *etc.*

Option 3

This looked promising at first sight as Exxon, ICI and BP all produced more ethylene than they needed and exported it. However, three of the sources are situated on the existing pipeline (Grangemouth, Mossmorran and Wilton) and would be subject to the same problems of supply. The other two sources are at Baglan Bay in South Wales and Fawley near Southampton (*Fig 3*) which make ethylene from imported oil. Ethylene would need to be transported from these locations to Stanlow and would therefore need other pipelines or alternative transport, as in Option 4.

Option 4

Shell considered supplying ethylene to Stanlow by ship, rail and road.

▼ **By ship**

The technology for this is already available. The ethylene must be transported as a liquid, *ie* at below 169 K. Facilities for cryogenic (low temperature) tankage are already available at Braefoot Bay on The Firth of Forth but 10 000 tonnes worth of tankage would have had to be built at Tranmere on the Mersey at a capital cost of £40 million. As well as the cost, the impact of this development on a highly populated area was considered. In addition, shipping costs were estimated at £3 million per year. The security of shipping was also considered as a negative factor because of its possible susceptibility to strikes, terrorists and accidents.

▼ **By rail**

Rail transport was discounted on similar grounds to shipping. A rail link to Mossmorran would have had to be built as well as storage facilities for liquid ethylene at Stanlow. The safety and environmental implications of moving large quantities of ethylene by rail were also factors.

▼ **By road**

Again, the technology is well known; six tankers per day already carry ethylene from Baglan Bay to Carrington – a total of 25,000 tonnes. To make up the projected shortfall, 60 would be needed. This would require capital expenditure on improving the loading and unloading facilities as well as running and maintenance costs for the vehicles. The safety and environmental considerations were similar to those for the rail option.

RS•C

Pipeline

Having rejected the other options, Shell looked at alternative pipeline options.

The existing trans-Pennine pipeline could only be uprated to a maximum throughput of 650,000 tonnes per year – insufficient to meet the needs of both Shell (610,000 tonnes per year and ICI (200,000 tonnes per year).

A pipeline from Baglan Bay would also produce insufficient ethylene – the maximum output of Baglan Bay is only 160,000 tonnes per year.

Selecting the route

Two quite different routes between Grangemouth and Stanlow/Carrington were considered – one of 360 km with a 160 km undersea section, and an approximately 400 km overland route (*Fig 4*). The partly undersea route was rejected on the basis of difficulties in construction, greater maintenance costs and the fact that valves could not be installed in the undersea sections. These factors outweighed the slightly shorter length.

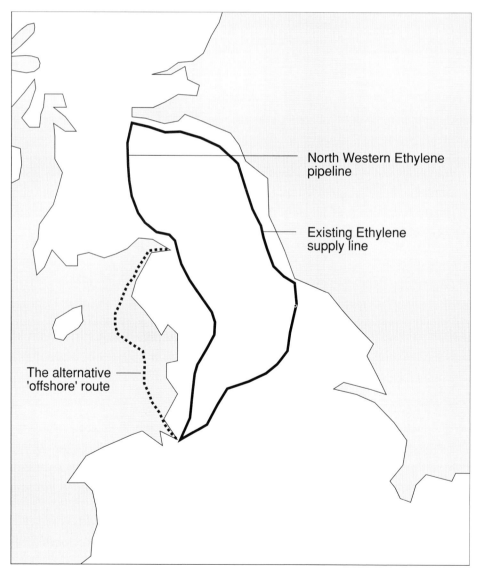

North Western Ethylene pipeline

Existing Ethylene supply line

The alternative 'offshore' route

Figure 4 Proposed pipeline routes

Overall, the most direct land-based route between Grangemouth and Stanlow was chosen. In detail, a great many factors had to be taken into account. These included safety considerations, land ownership, crossing existing facilities (roads, railways, waterways *etc*), geological conditions, historic, archaeological and scientific sites and beauty spots.

In general, the pipeline was routed away from centres of population for safety reasons. In densely populated areas, pipelines must be laid more deeply and have more protection because disturbance by digging is more likely in such areas.

Existing facilities such as motorways, roads, railways, canals, electricity power lines and existing pipelines tend to follow the same routes ("corridors") for geographical reasons (avoiding high land, major rivers *etc*). Crossing existing facilities (and rivers) was to be avoided as much as possible because of the cost of construction. Most of the pipe was to be laid in an open trench which was then filled in (the open cut method) but this is not possible when crossing main roads, railway lines and canals because of the disruption to traffic. These would have to be crossed by tunnelling or boring underneath – much more expensive. Regulations do not allow a pipeline to be laid within 6 m of an electric power line but, with this proviso, it makes sense to follow a similar route to existing facilities to take advantage of route preparation which has already been done (*Fig 5*).

RS•C

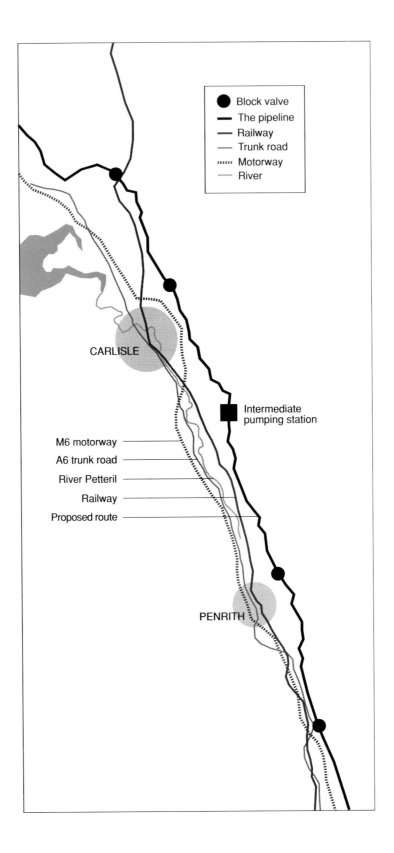

Figure 5 Section of proposed route showing how it follows existing features

RS•C

Eventually the route required the following crossings:

Motorways   8
A roads   47
B roads   33
Railways   21
Rivers   32
Canals   9

Woodland areas were to be avoided as much as possible on conservation grounds and because of the cost of felling trees. Archaeological sites were to be identified and either avoided, tunnelled beneath or professionally excavated prior to the pipe being laid.

The types of usage of land crossed by the pipeline are summarised below:

Cereal and rape   16.8%
Bare soil crops   7.2%
Grass   53.5%
Reverting grass   3.0%
New plantation   1.2%
Coniferous plantation   3.6%
Deciduous wood/scrub   1.1%
Heath/moor   11.8%
Recreation   0.1%
Other   1.7%

The overall route chosen runs southwards from Grangemouth to Carstairs and then parallel to the A74 to the Scottish border. It then follows the A74 and M6 through the Cumbrian mountains (east of the Lake District) and into Lancashire.

The whole of the pipeline is underground, with the excavated trench restored to its original condition. The following facilities are above ground (called AGIs or above ground installations) and are therefore the only parts visible.

▼   The terminal facilities at Grangemouth and Stanlow. These are within the existing industrial complexes in each case.

▼   The intermediate pumping station (Fig 6). This is situated near Carlisle next to a British Gas plant and is screened by trees.

▼   Two pigging stations (Fig 7) at Moffat and Garstang (the intermediate pumping station is also a pigging station). Pigs are devices which travel along the pipe with the gas flow, carrying sensors to monitor things such as signs of corrosion within the pipe. The pigging stations allow pigs to be inserted into and recovered from the pipeline.

▼   25 block valves (Fig 8). These are spaced at approximately 16 km intervals and are used to isolate sections of the pipeline in the case of a leak or for maintenance. The valves can be operated manually or by remote control from Stanlow.

RS•C

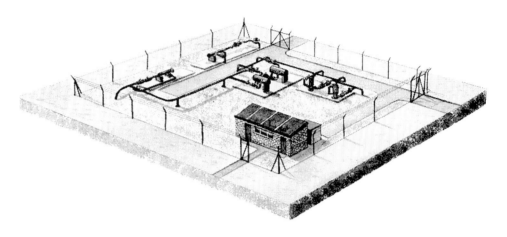

Figure 6 A pumping station

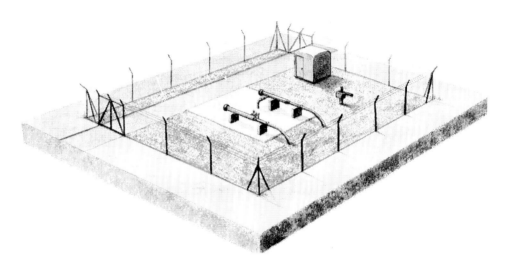

Figure 7 A pigging station

RS•C

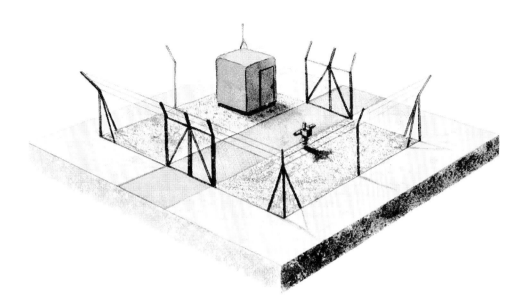

Figure 8 A block valve

The route in detail

Once the general route had been chosen, a 2 km wide corridor was surveyed including being shot on video tape from a helicopter. This corridor avoided problematic areas such as built-up areas and mountainous terrain. Records were then checked to identify any Sites of Special Scientific Interest (SSIs), environmentally sensitive areas, Scheduled Ancient Monuments and other archaeological sites. At the same time discussions were begun with Local Authorities, Water Authorities, relevant government departments and other interested parties such as the Nature Conservancy Council. On the basis of the original survey and these discussions, some changes to the route were made and the route corridor was narrowed to 500 m. The route was then walked from end to end.

The result of this was a 300 page statement on the environmental impact of the proposed pipeline. This was required to comply with UK and European Community legislation which requires the environmental impact of all major new construction projects to be assessed.

At this stage, between October and December 1989, the proposed route of the pipeline was publicly advertised. This involved supplying information to libraries and local planning offices along the route. The relevant MPs were also informed. This resulted in some 280 objections to the pipeline, most of which were withdrawn after discussions with Shell representatives. Nevertheless public enquiries were held at three venues close to the route. Landowners were offered compensation of £8.60 per metre of pipe crossing their land. Eventually a Pipeline Construction Authority (PCA) was granted in March 1991 and construction could begin.

Construction

Construction had to be completed within the March to October season for weather reasons. The task was split into four spreads – Grangemouth to Moffatt, Moffatt to Penrith, Penrith to Cockerham and Cockerham to Stanlow each constructed simultaneously by a separate sub-contractor (*Fig 9*). These laid an average of 500 m of pipe per day.

RS•C

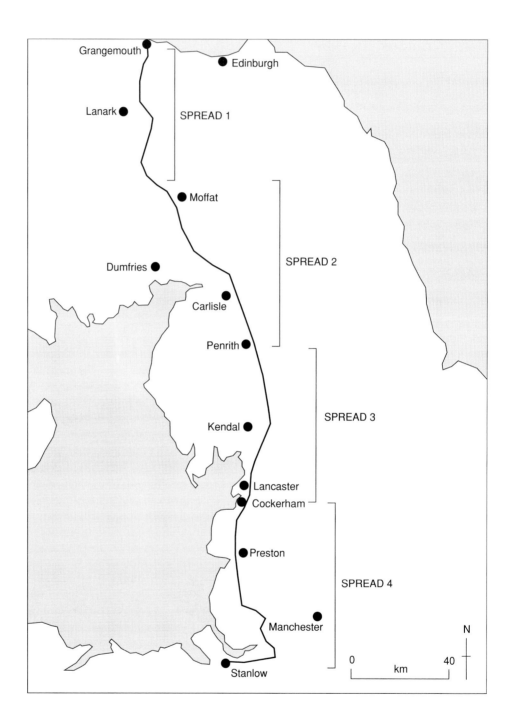

Figure 9 Construction spreads

The pipe itself is 10" (254 mm) in diameter and comes in over 30,000 13.65 m steel sections which are welded together on site. The pipe is coated with epoxy resin to protect against corrosion. Pipe lengths were stored in thirteen pipe yards spread out over the route each covering around 4 acres (16,000 m²). On completion of the line, these were dismantled and the ground restored to its original condition.

RS•C

Standard construction

The standard method of pipe laying (*Fig 10*) involves first fencing off a strip of land 20 m wide, called the working width. Within this strip the topsoil is removed and stored. In sensitive areas, turf is removed and marked so that it can be replaced in exactly the same place. Hedges and fences are removed but, where possible, trees are left in place. The sections of pipe are then brought to the site, welded together and the welds tested by X-ray or ultrasound. The pipe is then epoxy coated. A trench 1 m wide and 3 m deep is next excavated and the subsoil is stored separately from the topsoil. The pipe is then lowered into the trench by teams of side boom tractors. As each tractor lowers a section of pipe, it moves to the front of the line to lift the next section. The trench is then backfilled with about 15 cm of subsoil to protect the pipe's coating. The interior of the pipe is cleaned using a pig and a second pig is passed through to check the interior diameter to ensure that no deformation has occurred. The line is next tested by filling with water under pressure for 24 hours. The trench is then restored by filling with the rest of the subsoil and the topsoil, taking care to reconnect any land drains. The working width is then ploughed to reduce compaction of the soil caused by the passage of heavy vehicles. Finally hedges, walls and fences are replaced. The aim is that eventually the line of the pipe will become undetectable, except for markers erected to show its position.

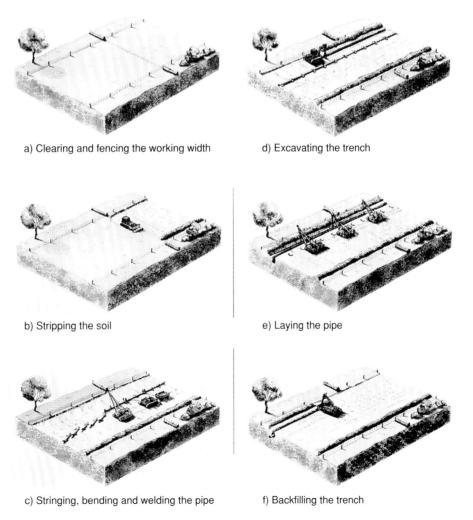

a) Clearing and fencing the working width   d) Excavating the trench

b) Stripping the soil   e) Laying the pipe

c) Stringing, bending and welding the pipe   f) Backfilling the trench

Figure 10 Stages in laying the pipe

RS•C

Special construction

For some road, rail, river and canal crossings, the open cut method of pipe laying is not feasible and special techniques are used to drill beneath the obstruction.

Auger boring and thrust boring are similar to one another in that pits are dug on either side of the obstruction. Drilling equipment is placed in one pit to drill through to the second, reception, pit.

Directional drilling is a technique borrowed from the oil industry and avoids digging the pits at either side of the obstruction (*Fig 11*).

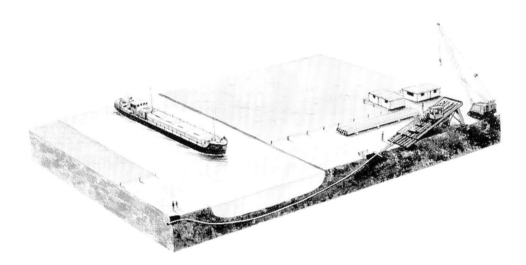

Figure 11 Directional drilling

Safety and maintenance

Corrosion

The pipe is coated with epoxy resin and in addition is protected from corrosion by a cathodic protection system. This involves establishing a potential difference between the pipeline and an inert electrode buried close to the pipe such that the pipe is the cathode (negative electrode) (*Fig 12*) . The excess of electrons on the steel pipe inhibits the formation of Fe^{3+} ions by the half equation

$$Fe \rightleftharpoons Fe^{3+} + 3e^-$$

effectively forcing this equilibrium to the left.

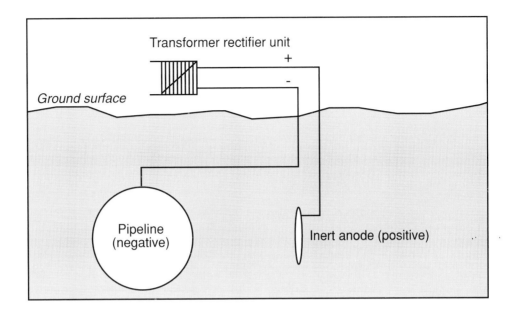

Figure 12 Cathodic protection

Cathodic protection points are situated every 32 km.

Question 4
An alternative method of cathodic protection is used to protect undersea pipelines. This involves welding lumps of zinc to the pipeline at intervals. How does the zinc provide electrons to displace the Fe/Fe^{3+} equilibrium? Discuss pros and cons of the two systems for undersea and underground pipelines.

Monitoring and maintenance

The pipe requires little routine maintenance.

The route is inspected fortnightly by helicopter and is walked once a year to locate any changes which might affect the pipe. Block valves are inspected monthly.

A variety of pigs can be sent down the line from the three pigging stations. Scraper pigs can be used to remove any coating of polymerised ethylene which may form on the inside of the line and calliper pigs can detect any changes of internal dimensions and so warn of buckling.

Cost

Originally projected at £90 million, the project eventually cost £150 million. It is worth considering this figure in the light of the projected figures for transporting the ethylene by ship – construction of tankage at a capital cost of £40 million and annual operating costs of £3 million per year.

RS•C

Statistics

Pipeline length	411 km
Cost	£150 million
Construction employment	2000–2500 (at peak)
Construction timescale	May–December 1991
Testing timescale	December 1991 – April 1992
Commissioning	June 1992
Parties negotiated with	881 (824 landowners and tenants, 32 local authorities, 25 environmental bodies)
Plots of land crossed	1348

RS•C

Answers to questions

1. Ethane-1,2-diol: $CH_2(OH)CH_2(OH)$

    Propenenitrile: CH_3CH_2CN

    1,2-Dichloroethane: CH_2ClCH_2Cl

    Ethene: $CH_2=CH_2$

    Ethanol: CH_3CH_2OH

    Ethanal: CH_3CHO

    Ethanoic acid: CH_3COOH

    Ethanoic anhydride: $CH_3COOCOCH_3$

    Ethylbenzene: $C_6H_5CH_2CH_3$

    Phenylethene: $C_6H_5CH=CH_2$

    Chloroethane: CH_2ClCH_3

    1,2-Dichloroethane: CH_2ClCH_2Cl

    1,2-Diaminoethane: $CH_2(NH_2)CH_2(NH_2)$

    Chloroethene: $CHCl=CH_2$

    1,2-Dibromoethane: CH_2BrCH_2Br

2. $CH_3CH_2CH_2CH_2CH_3 \rightarrow CH_2= CH_2 + CH_3CH_2CH_3$

    $CH_3CH_3 \rightarrow H_2 + CH_2= CH_2$

3. $CH_4(g) + 2O_2(g) \rightarrow 2H_2O(l) + CO_2(g)$

    $C_2H_6(g) + 3\frac{1}{2}O_2(g) \rightarrow 3H_2O(l) + 2CO_2(g)$

    The two fuels require different ratios of fuel to oxygen and would therfore require different burner designs.

    Ethane can be cracked to ethene, the starting material for all the products in *Fig 2*. Methane, with only one carbon, cannot. Ethane is therefore too valuable to be burnt as a fuel.

4. Zinc is more reactive than iron so the zinc will corrode faster

    $Zn \rightleftharpoons Zn^{2+} + 2e^-$

    As the zinc is in contact with the iron, these electrons force the Fe/Fe^{3+} equilibrium to the left as in the electrical method.

    Considerations include:

    ▼   lack of electrical power sources undersea and the problem of installing them;

    ▼   the capital cost of the zinc blocks compared with the running cost of the electricity; and

    ▼   the zinc lumps will eventually dissolve away (in fact such anodes have a design lifetime of about 25 years).

RS•C

Steelmaking in the UK

British Steel plc has taken significant steps to reduce emissions at its plants to comply with legislation and in response to public concern about the environment, both local and global. This is helping to change the image of iron and steel manufacture which has traditionally been thought of as dirty and smelly with plant producing dust, smoke and unpleasant fumes. This study looks at emission control schemes at:

▼   coke ovens at Scunthorpe, North Lincolnshire;

▼   a modern blast furnace which produces iron from iron ore;

▼   a new slag granulator at Llanwern, South Wales;

▼   the iron desulfurisation plant at Lackenby near Redcar on Teesside; and

▼   an oil fume catalytic oxidiser at Cookley in the West Midlands where coated steel for vehicle petrol tanks and gas meters is produced .

Figure 1 Selected British Steel plc sites

RS•C

The legislative background   **Box 1**

The main legislation in environmental control is the Environmental Protection Act (EPA), 1990 which implemented Integrated Pollution Control (IPC). More recently, the 1995 Environment Act has set up the Environment Agency which has taken over the functions of a variety of organisations such as Her Majesty's Inspectorate of Pollution (HMIP), The National Rivers Authority (NRA) and waste regulatory authorities.

The principles of IPC are:

▼   to prevent or minimise the release of prescribed substances (defined in separate lists) and to render harmless any such substances which are released; and

▼   to develop an approach to pollution control that considers releases from industrial processes to all media (*ie* air, land and water) in the context of the effect on the environment as a whole – the Best Practicable Environmental Option (BPEO).

Within this framework the Environment Agency or The Secretary of State for the Environment can set limits for releases of substances. It is recognised that costs (both capital and running) must be considered and authorisations are agreed using the principle of Best Available Techniques Not Entailing Excessive Costs (BATNEEC). This means that the emissions from an existing plant might be allowed to be greater than those from a new one because the new plant could have the tighter emission limits designed in whereas it could be prohibitively expensive to modify an existing system.

RS•C

Steelmaking – a brief overview

Steels are alloys of iron with carbon (typically 0.1 to 0.3%) and other metals such as chromium, manganese and vanadium. Different types of steel have different properties – hardness, toughness and corrosion resistance for example.

Iron and steel making from primary raw materials (coal, iron ores and limestone, *ie* those dug from the ground) has six stages.

▼   Converting coal into coke for use in the blast furnace;

▼   converting iron ores into sinter (Box 3) for use in the blast furnace;

▼   making iron from sinter and coke in the blast furnace;

▼   desulfurising the iron to remove sulfur originating largely from the coke used in the blast furnace;

▼   primary steelmaking, in which oxygen is blown through molten iron from the blast furnace to reduce the percentage of carbon and remove non-metallic impurities; and

▼   secondary steelmaking in which further changes in composition are achieved.

Coke-making

Coal is not (as is often assumed) pure carbon. Coals from different sources have different compositions, all of them having a complex macromolecular structure. As well as carbon they contain hydrogen and oxygen and also smaller quantities of sulfur, nitrogen and some trace elements. In coke-making, a blend of coals is heated in the absence of air. The coal decomposes, volatile compounds are driven off and carbonisation takes place leaving a porous solid containing around 88% carbon but with significant amounts of impurities including around 0.6% sulfur.

Latex spraying   **Box 2**

The steelworks on Teesside covers an area of about 12 km² by the Tees estuary. Iron ores, limestone and coal arrive by ship and are unloaded and stored in the open, prior to being blended and processed. Wind can blow away quantities of these materials causing loss to British Steel and an environmental nuisance to those downwind. The piles are now sprayed with a latex solution which sets and coats the piles to prevent wind lift off.

The efficiency of the system was illustrated one day when a problem developed with the latex in the storage tank and for a period the stock piles were sprayed with water only; a number of complaints was received from residents of nearby Redcar.

RS•C

Ironmaking

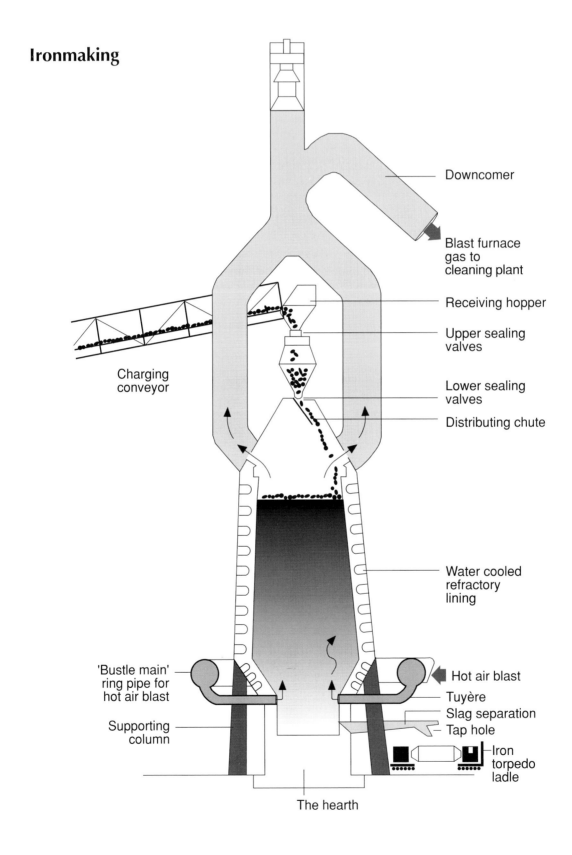

Figure 2 Blast furnace

RS•C

A blend of finely crushed iron ores from different sources, containing the iron oxides haematite (Fe_2O_3) and magnetite (Fe_3O_4), as well as silica-based impurities, is mixed with crushed coke (largely carbon) and limestone (mostly calcium carbonate). This mixture (called the charge) is heated on a moving grate in a process called sintering and converted into a sintered (porous) material (see *Box 3*) which is fed into the blast furnace (*Fig 2*) together with lump iron ore, coke and other materials. The blast furnace is a tower around 40 m high, made of steel, lined with refractory (heat-resistant) material which is water-cooled. Pre-heated air is blown into the base of the furnace. Sintering is necessary to produce a hard, porous material called clinker which will allow the passage of air through the contents of the furnace (called the burden) and to give the burden strength to stop it collapsing under its own weight.

Sintering   Box 3

The natural raw materials for the blast furnace are of variable quality depending on their sources. To feed the blast furnace with a consistent material, the raw materials – fine coke, iron ores and limestone – are blended using a boom stacker which places up to 15 different materials in horizontal layers of appropriate thickness to give the correct blend on an 800 m long ore bed.

Photo: British Steel SP&CS Scunthorpe Works

Boom Stacker making an ore bed

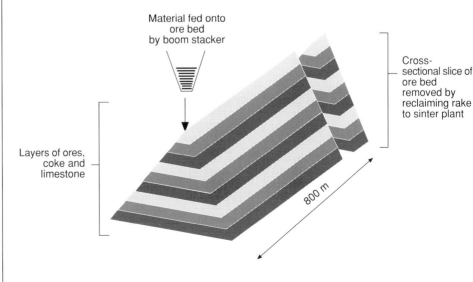

Material fed onto ore bed by boom stacker

Cross-sectional slice of ore bed removed by reclaiming rake to sinter plant

Layers of ores, coke and limestone

800 m

A schematic representation of an ore bed

RS•C

> **Box 3 continued**
>
> Recycled iron-bearing materials are also added at this stage. A reclaiming rake removes a cross-section of this pile of layers and carries it to the sinter plant. Here the mixture is ignited. This melts the surface of the particles so that they fuse together. The resulting sinter is screened for size before being fed into the blast furnace.
>
>
>
> **A reclaiming rake working on an ore bed**

The main reactions which occur in the blast furnace are:

$CaCO_3(s) \rightarrow CaO(s) + CO_2(g)$

$Fe_2O_3(s) + 3CO(g) \rightarrow 2Fe(l) + 3CO_2(g)$

$CaO(s) + SiO_2(s) \rightarrow CaSiO_3(l)$

$CO_2(g) + C(s) \rightarrow 2CO(g)$

$Fe_2O_3(s) + 3C(s) \rightarrow 2Fe(l) + 3CO(g)$

$C(s) + O_2(g) \rightarrow CO_2(g)$

Question 1
The above equations are written for haematite, Fe_2O_3. Write balanced equations for the corresponding reactions with magnetite, Fe_3O_4.

Question 2
Four of the reactions above are redox reactions. Use oxidation numbers to identify these. What is being oxidised and what is being reduced in each case?

Liquid iron and slag (primarily calcium silicate ($CaSiO_3$) descend to the base of the furnace (called the hearth). The slag floats on top of the molten iron and the two liquids can be tapped off and separated. A mixture of gases containing carbon monoxide is produced. This is cleaned and used as a fuel in other parts of the steelmaking process. Slag (which locks up the silica-based impurities in the ore and ash from the coke) is cooled and solidifies to a greyish solid. Slag was once stored in unsightly slag heaps but is now fully used for road building and in cement

RS•C

manufacture. Some of the sulfur present in the coke is incorporated into the slag as calcium sulfate and calcium sulfide and some sulfur remains in the iron. During the water-cooling process, calcium sulfide can react to give hydrogen sulfide producing the well-known bad eggs smell.

$CaS(s) + 2H_2O(l) \rightarrow Ca(OH)_2(s) + H_2S(g)$

Question 3

What is the balanced equation for the combustion of carbon monoxide? Use the values of ΔH_f° below to work out ΔH° for this reaction.

$\Delta H_f^\circ CO_2 = -393.5 \text{ kJ mol}^{-1}$

$\Delta H_f^\circ CO = -110.5 \text{ kJ mol}^{-1}$

Desulfurisation

British Steel's blast furnace at Redcar on Teesside has a design capacity of 10,000 tonnes of iron per day – as much as was produced by 100 blast furnaces at the turn of the century. It operates continuously, 24 hours a day, 365 days a year, with iron being tapped in turn from each of four tapholes at its base. The furnace was built in the 1970s and has a current campaign life until 2005 at which time the furnace's lining of refractory brick will be replaced.

Iron from this blast furnace contains around 0.02% of sulfur (chemically combined as iron sulfide, not free). This percentage must be reduced, before steelmaking, to below 0.012% because high sulfur levels can result in cracking of the steel during continuous casting and surface defects and poor weldability in the finished steel. There are a number of ways of desulfurising iron including injecting one of a variety of substances into the molten iron. These include calcium carbide, lime (calcium oxide), soda ash (sodium carbonate) and magnesium metal.

Question 4

What are the possible products of the reaction of each of the four reagents above with sulfur?

In the calcium carbide method used at Redcar, a mixture of calcium carbide and calcium carbonate is blown in a stream of nitrogen gas into the molten iron. The function of the calcium carbonate is to produce bubbles of carbon dioxide as it decomposes thermally in the hot, molten iron which is at approximately 1500 °C. These bubbles promote mixing. The resulting slag, containing the sulfur as calcium sulfate and calcium sulfide, is removed from the surface of the molten iron mechanically, a process called rabbling.

Question 5

Use the equation $\Delta G^\circ = \Delta H^\circ - T\Delta S^\circ_{system}$ to show that the decomposition of calcium carbonate is a feasible process at 1500 °C.

$\Delta H_f^\circ CO_2 = -393.5 \text{ kJ mol}^{-1}$	$S^\circ CO_2 = 213.6 \text{ J K}^{-1} \text{mol}^{-1}$
$\Delta H_f^\circ CaCO_3 = -1206.9 \text{ kJ mol}^{-1}$	$S^\circ CaCO_3 = 92.9 \text{ J K}^{-1} \text{mol}^{-1}$
$\Delta H_f^\circ CaO = -635.1 \text{ kJ mol}^{-1}$	$S^\circ CaO = 39.7 \text{ J K}^{-1} \text{mol}^{-1}$

What assumption are you making in this calculation?

RS•C

Primary basic oxygen steelmaking

After desulfurisation, the iron still contains a number of impurities including carbon, silicon and phosphorus. These are removed in the basic oxygen steelmaking process (BOS) by blowing a stream of pure oxygen through the molten iron and adding lime (calcium oxide, CaO). Non-metallic impurities are converted to their acidic oxides and these react with the lime (which is basic) to form steelmaking slag which is separated from the steel during tapping. Most of the carbon in the iron is released as carbon monoxide which can be used as a fuel on site.

For example

$$4P(s) + 5O_2(s) \rightarrow 2P_2O_5(s)$$

$$P_2O_5(s) + 3CaO(s) \rightarrow Ca_3(PO_4)_2(s)$$

This slag can be spread on farmland as it contains unreacted lime and also supplies essential elements such as phosphorus and sulfur to the soil.

All these reactions are exothermic. Up to 25% solid steel scrap is added at this stage to be recycled and helps to cool down the steel. At this stage, too, small quantities of other elements are added to make steels with particular properties.

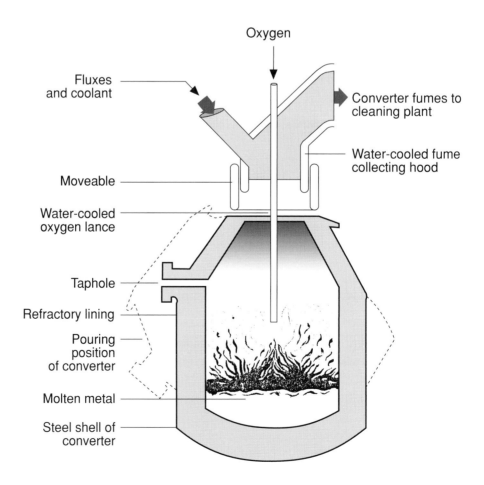

Figure 3 Basic oxygen steelmaking

RS•C

Secondary steelmaking

The steel resulting from primary steelmaking contains more than the desired amount of oxygen, both as bubbles of gaseous oxygen and as iron oxide. Aluminium, silicon or manganese are commonly added to the steel in the secondary steelmaking process and these react with this oxygen to form oxides which float on top of the molten steel and can be removed. Further additions are often made to produce steels of particular specifications.

RS•C

Environmental issues

Dawes Lane coke ovens at Scunthorpe

Blast furnace coke is made by heating coal for at least 18 hours at about 1100 °C in a battery of gas heated ovens. Dawes Lane has 75 ovens side by side. Each oven is 5.4 m high, 15 m long and only 0.5 m wide. The narrowness is to allow good heat transfer from the walls of the oven to the centre of the coal. The battery of ovens is constructed from refractory (heat resistant) silica-based bricks within a steel frame. Each oven is separated from the next one by a gas-heated chamber in such a way that the battery of ovens forms a multi-layered sandwich. Each oven has a steel door on each side and four charging holes on top (*Fig 4*).

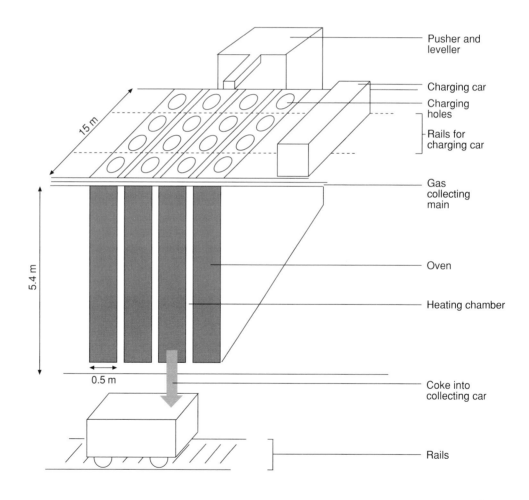

Figure 4 (a) Part of a coke oven battery viewed from the coke side

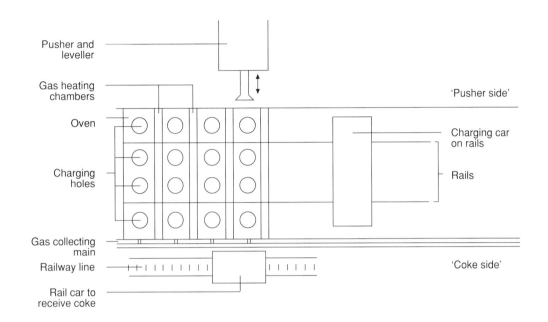

Figure 4(b) General layout of coke oven – top view

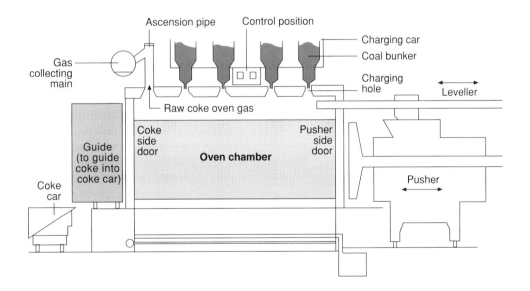

Figure 4(c) General layout of coke oven – end view

Coal is imported from countries such as Canada, Poland, Australia and the US. (British coal is generally of lower quality and can have too high a sulfur content for coke making, so only small quantities can be used.) Coal stockpiles are sprayed with a crusting agent to prevent coal being blown away in high winds which would cause loss of coal and a dust nuisance. Different coals are first blended, to give a mixture with the correct proportion of volatile material (about 24%), and then crushed so that the majority of particles is smaller than 3 mm. A small amount of oil is added to help it flow smoothly through the charging equipment and to increase its bulk density.

Coal is fed into the ovens from a charging car which runs on a railway track above the ovens. The car has four hoppers each holding about seven tonnes of coal. These feed a total charge of 28 tonnes of coal into an oven through four charging holes in the top of each oven.

The charge is then levelled by a hydraulically operated levelling bar, which is inserted through a leveller door, to ensure the optimum conditions for carbonisation and a headspace above the coal to allow gases to escape (*Fig 5*).

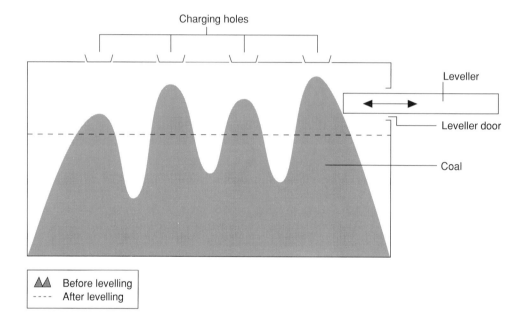

Figure 5 Levelling the coal

Over the 18 hour carbonisation period, moisture is first driven out of the coal followed by a gaseous mixture of volatile compounds, consisting of coke oven gas, crude tar, benzole and ammonia. This raw gas is collected at 800 °C from the ovens in a collecting main (a large diameter pipe running horizontally along the oven battery above the level of the ovens).

Initial separation of the components of this gas is achieved by cooling it to 80 °C by spraying with water to condense out the less volatile components of the mixture. The spray yields a dilute solution of ammonia called ammoniacal liquor. Crude tar is a complex mixture of organic compounds including many aromatics such as naphthalene. Benzole is a mixture of benzene, methylbenzene (toluene) and dimethylbenzenes (xylenes). The tar, benzole and ammonia by-products are sold, although sometimes ammonia is burnt if there is a lack of demand.

After the less volatile components have been separated, coke oven gas remains. This consists of a mixture of hydrogen, methane, carbon monoxide and nitrogen with a little hydrogen sulfide. The gas is cleaned and used as a fuel to heat the ovens and elsewhere in the works. Any gas which is excess to requirements is burnt at a stack (although for economic reasons this is minimised).

One tonne of coal produces 0.76 tonnes of coke, 0.025 tonnes of tar, 0.086 tonnes of benzole and 0.0073 tonnes of concentrated ammoniacal liquor. The gas produced has a total energy value of 5.95 GJ (gigajoules) which is more than the energy required to make a tonne of coke.

RS•C

Question 6

What percentage volume reduction would be brought about by cooling a gas from 800 °C to 80 °C (at constant pressure)? Assume ideal gas behaviour.

Question 7

What is the equation for the combustion of ammonia?

Question 8

$1m^3$ of coke oven gas contains 600 dm^3 of hydrogen, 240 dm^3 of methane and 70 dm^3 of carbon monoxide (the rest is largely nitrogen and carbon dioxide). Use the value of $\Delta H_c^\ominus$ for carbon monoxide calculated in question 3 and the values of $\Delta H_c^\ominus$ for hydrogen of -572 kJ mol^{-1} of H_2 and methane of -890 kJ mol^{-1} to work out the energy produced by burning 1 m^3 of coke oven gas. Assume the gas is at room temperature (298 K) and the volume of a mole of gas is 24 dm^3.

Coke ovens must be operated continuously, 24 hours a day, 365 days a year. If the battery of ovens is allowed to cool, the steel frames and silica brickwork can be irreparably damaged by the stresses caused by contraction.

After 18 hours of heating, the red-hot coke (about 90% carbon and 10% inorganic ash) is pushed. Doors at either side of the oven are opened and a hydraulic ram pushes the coke from the oven into a rail car which takes it to a wharf for water quenching. It then goes by conveyor to the blast furnace, sinter plant or to a stockpile. The empty oven is recharged with coal within 10 minutes of being pushed.

During the coke making process there is potential for a number of releases to the air:

▼   release of smoke and gas during charging;

▼   release of gas and smoke through the leveller door during levelling of the charge;

▼   leaks of gas and smoke from the oven doors and charging holes during carbonisation if these are not effectively sealed ; and

▼   release of smoke from the oven doors during pushing.

Note: smoke may include coal dust and tar which condenses as it cools on exposure to the cold air.

Various design features and operational procedures have been developed to minimise these releases.

Charging hole lids, which are similar to manhole covers, are removed and replaced automatically for charging by equipment on the charging car. When they are replaced, a dilute solution of gypsum (calcium sulfate) is dripped into the joint between cover and hole. The heat of the oven rapidly evaporates the water, sealing the hole with solid gypsum (plaster of Paris). Any few remaining leaks can be sealed by manual application of the solution.

During charging, the coal displaces a volume of gas equal to the volume of coal added. This cannot escape through the charging holes as these are sealed to the hoppers of the charging car and some coal is retained in the hopper preventing the displaced gas venting through it. The gas vents into the collecting main via the ascension pipes rather than being discharged to the air (Fig 4c).

During levelling, some coke oven gas can escape, but this gas ignites on contact with air, burning to give innocuous carbon dioxide and water.

RS•C

Oven doors are closed by knife edge metal-to-metal seals. Small leaks tend to seal themselves as tar from the coking process deposits around any small cracks. When the doors are opened for pushing, the seals are automatically cleaned by high pressure jets of water to remove any excessive build up of tar which might prevent effective sealing when they are closed. Larger leaks are more problematical and can only be effectively cured by replacing the door as each door fits only its parent oven. Emissions during charging and pushing are estimated visually on a five point scale defined by the British Carbonisation Research Association (BCRA) by trained operators whose scores are calibrated annually in accordance with Environment Agency requirements. The scores are used to calculate a percentage leakage control factor. The targets of 98% (doors) and 99% (charging holes) are normally achieved.

There are emissions arising when the oven doors are opened for pushing but if the coal has been properly carbonised, these should be minimal.

The modern blast furnace

It is not possible to give an accurate generic description of a blast furnace as each one is individually designed and built. However, a typical modern blast furnace produces 5,000 – 10,000 tonnes of molten iron (hot metal) per day. It is charged with layers of sinter and coke (the burden) via a bell-less top. This allows charging to take place with the minimum release of gases from the furnace to the atmosphere. This is important both from the point of view of reducing environmentally unpleasant gas emissions (the furnace gas is dusty and contains carbon monoxide) and because the gas is used as a fuel. The bell-less top also includes a steerable chute (computer controlled) which enables the burden to be placed exactly where required in the furnace leading to a smoother and more efficient process. The bell-less top leads to better sealing of the furnace top because the seals are not worn by contact with the burden on loading. Also there is less emission of furnace gas on charging (Box 4).

The pre-heated air blown into the base of the furnace is injected with coal or oil (some of which may be waste oil from the rest of the plant). This acts as a fuel to further heat the air and also produces hydrogen which assists the reduction of the iron ores. It also reduces the need for coke fuel and can be used to get rid of waste oil whose disposal would otherwise have to be paid for. The use of a blast furnace as an incinerator to burn other waste products has been considered, but furnace managers are often reluctant to investigate this for fear of destabilising the main process of iron making.

Typically, the molten iron is tapped (cast) from the furnace 10 to 20 times per day with 15 to 20 minutes between casts. Casting involves drilling through a clay-based plug which seals the tapping hole. Nowadays iron and slag are tapped through the same hole and separated by skimming (Fig 6). The iron is poured into rail cars called torpedo ladles which typically hold 300 tonnes each and carry the hot metal to the steelmaking plant.

Charging the blast furnace   Box 4

The earliest blast furnaces were charged manually through an open top. This meant that the furnace gases escaped and simply burnt off. Later a double bell system was used in which a cone-shaped bell blocks the top of the furnace, diverting the furnace gas into a pipe called a downcomer. A second, smaller, bell is placed above the first one – see below. The charge is placed on the upper bell which is then lowered allowing the charge to fall onto the lower one. The upper bell is then raised, sealing the top of the furnace and lower bell dropped allowing the charge to fall into the furnace. The only gas which escapes is that which fills the space between the bells. Modern furnaces use a system of hoppers and sealing valves and the charge is placed in the required position in the furnace by a computer-controlled chute.

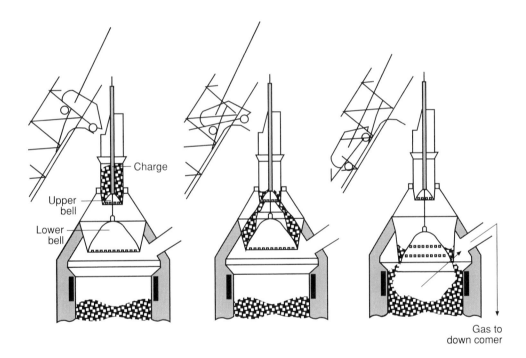

Stages in charging a blast furnace via a double bell system

Monitoring the blast furnace   Box 5

It is not, of course, possible to see inside a blast furnace. Thermocouples can be used to measure the temperature at different points. Gas analysis and stock level probes can also be inserted at various positions. Together, these allow the furnace manager to monitor the condition of the furnace interior.

Much of what is known about the blast furnace's chemistry has come from examination of a number of Japanese furnaces which were ready to be relined. They were quenched with water to stop the reactions and were then "dismantled" bit by bit to determine the physical and chemical conditions at various points. A German furnace was investigated similarly after quenching with nitrogen.

RS•C

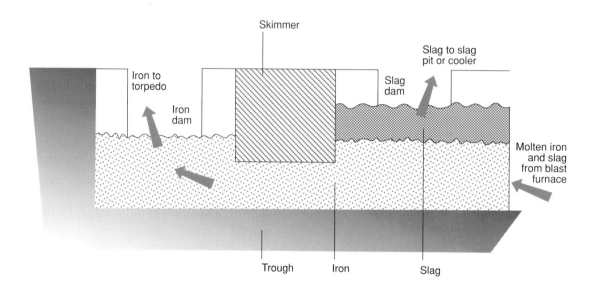

Figure 6 Slag skimming on tapping the blast furnace (side view)

On tapping, the hot metal is exposed to air and it oxidises, producing brown fumes of iron(III) oxide. The more turbulent the flow, the more fumes are produced. How turbulent the flow is depends to some extent on the character of the taphole and can not always be predicted exactly. The brown fumes are visually unpleasant and are likely to fall out over nearby areas. Iron (III) oxide is, of course, a raw material for the process and therefore has a value. Modern furnaces have fume hoods to collect the fumes and carry them to a bag filtration plant which is similar to the one described in the section on desulfurisation (p. G22). This collects about 99% of the dust and emits air with less than 50 mg m^{-3} of dust. Several thousand tonnes of iron oxide dust (worth about £12 per tonne, 1996 prices) is recycled to the sinter plant each year. A futher benefit of the fume extraction system is operator comfort; it can be impossible to see across the cast house of a blast furnace without fume extraction.

Plating

Blast furnaces run continuously thoughout their lifetimes – a typical campaign might be 12 years before the refractory linings become so worn that they need to be replaced. However, in some circumstances, such as a breakdown in the steel making plant, there may be no torpedo ladle capacity spare to hold the hot metal. In these cases it is necessary to pour the molten iron into pits where it solidifies. This process is called plating and can also cause brown fumes. This is minimised by the design of the pits to prevent splashing and, in some cases, the use of nitrogen to displace air (a technique known as suppression).

Cleaning the furnace gases

The gas produced by the ironmaking process has the approximate composition: CO, 21%; CO_2, 21%; H_2, 4%; N_2, 54%. It also contains about 2 g m^{-3} of dust before cleaning.

Question 9
Which two of these gases are fuels?

RS•C

This gas passes first through a dust catcher – a U-shaped bend. As the gas slows down at the base of the U, most of the larger particles of dust simply settle out. The smaller dust particles are then removed by water sprays in a scrubber unit. This water is passed to a clarifier where the solid particles settle out and the cleaned water is re-used in the plant. The sludge is removed, dried and eventually recycled through the sinter plant.

The treatment of ironmaking slag

Each tonne of iron produced in the blast furnace also produces about 0.25 tonnes of slag. At one time this was regarded as a waste material and discarded in unsightly slag heaps. Nowadays, slag is a saleable by-product with uses as roadstone and in the manufacture of cement. There are three methods of treating slag.

1. Air cooling

Here the molten slag from the blast furnace is run off into pits (120 tonnes at a time) and allowed to cool and solidify (*Fig 7*). The successive charges of slag are separated by horizontal weaknesses in the structure. Although referred to as air cooling, the cooling process is accelerated by spraying water onto the hot slag which causes vertical cracks. The horizontal and vertical cracks make it relatively easy to quarry the slag from the pits. It is then mechanically crushed to fractions of various sizes and can be used as stone for road building and as filter media in sewage treatment plants. While the liquid slag is cooling, sulfur dioxide is given off and, when the slag is sprayed with water, hydrogen sulfide can be produced. Both these gases are potential pollutants because of their smell and toxicity: sulfur dioxide contributes to acid rain. Some work has been done to try to suppress hydrogen sulfide formation by adding lime to the cooling water with considerable success at one site.

Photo: Skyscene

Figure 7 Slag cooling in slag pits

RS•C

Question 10

a)   Suggest an equation for the formation of hydrogen sulfide from calcium sulfide during the water cooling of slag.

b)   Classify the chemical behaviour of the water in this reaction as either: oxidising agent or reducing agent or acid or base or none of these.

c)   Suggest how the addition of lime (calcium oxide) might help to suppress the reaction.

Question 11

a)   Write an equation for the conversion of sulfur dioxide into sulfuric acid in the atmosphere.

b)   What two other reactants are required?

2. Pelletising

In pelletising, the molten slag is fed into a rotating, slotted drum. On emerging from the slots, it meets a stream of cooling water which causes it to solidify into pellets (up to 20 mm in size). These can then be ground into powder suitable for making cement which is called (unsurprisingly) ground pelletised blast furnace slag. It is worth appreciating that the process of manufacturing cement by heating clay (an aluminosilicate) with limestone (calcium and magnesium carbonates) is very similar to the process of slag formation in the blast furnace so it is not surprising that the products are similar.

3. Granulating

This is the most recent process and produces least air pollution. The slag from the blast furnace enters a hopper where it meets high pressure water at a ratio of 8–10 tonnes of water to 1 tonne of slag (*Fig 8*). The slag is cooled almost instantaneously, remaining in a vitrified (glassy, rather than crystalline) state and looking rather like coarse sand. The hot slag is not left exposed to air so sulfur dioxide is not evolved and any sulfur is locked in to the vitrified material. The cooling water is separated and recycled via a cooling tower, losses by evaporation being made up with water which has been used to clean the blast furnace gases. The granulated product is sold on to another company which grinds it to powder. This is used mainly in the production of concrete in which it is mixed with Portland cement. It speeds up the chemical reactions which occur when concrete sets. The resulting concrete is more durable than straight Portland cement.

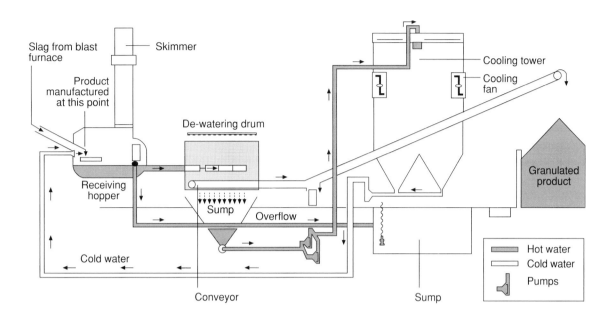

Figure 8 Cambrian Stone Llanwern slag granulation plant

At the Llanwern site in South Wales the slag is treated by Cambrian Stone, a company jointly owned by Tarmac Quarry Products and British Steel plc. The company supplies limestone to British Steel for its blast furnaces and also processes the slag – dealing with 2.5 million tonnes of material per year. Slag is dealt with by both the aircooled method and by granulation (*Fig 9*), the relative amounts depending on the commercial demand for the different products.

The second Severn crossing under construction   **Box 6**

Photo: Skyscene

This bridge over the River Severn from England to South Wales (opened in 1996) used a significant amount of blast furnace slag in its construction, much of it coming from the British Steel plant at Llanwern a few miles away. Slag was used for making the roads and also in the cement used in the concrete of the bridge itself.

RS•C

Photo: Skyscene

Figure 9 Slag pit (centre) with the granulation plant on the right

Desulfurisation at Redcar

At Redcar, molten iron (known as hot metal) is carried from the blast furnace by rail a few km across the site to the steelmaking plant. It is carried in torpedo ladles, so-called because of their shape, which hold up to 300 tonnes of hot metal. Traditionally the carbide injection (see p. G7) was carried out in the torpedo ladle before transfer of the molten iron into the steelmaking vessel (*Fig 10*). On injection of the powdered calcium carbide and calcium carbonate mixture through a lance into the hot metal, there was considerable splashing of molten metal from the torpedo and onto the ground. This exposed molten iron droplets to the air and these were rapidly oxidised to iron(III) oxide, producing copious brown fumes. These were highly visible; they caused many complaints from the public and affected the working environment. Fine, sparkling, black platelets of graphite (called kish) were also produced as the carbon which was dissolved in the steel precipitated out on cooling. Under this system, the process was carried out under a fume collection hood with a fan which drew the oxide dust into a chimney.

Question 12

Why do splashed droplets of hot iron produce more brown fumes than the same amount of undisturbed molten iron at the same temperature.

RS•C

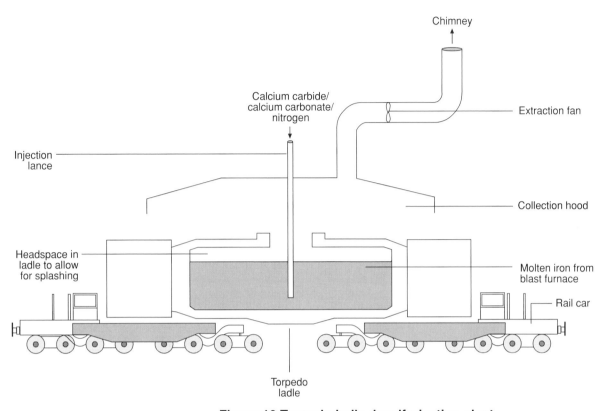

Figure 10 Torpedo ladle desulfurisation plant

By the 1980s, demand for lower sulfur steels increased and a new desulfurising facility was required. This was installed in 1992 at a capital cost of £7.5 million (one third of which was for environmental control equipment). Under the new system, the molten iron is first transferred from the torpedo into a transfer ladle for desulfurisation and then into a steelmaking vessel. This allows desulfurisation and slag skimming to be carried out in an enclosed chamber from which the brown iron oxide fumes can be collected for cleaning via a bag filtration system (*Fig 11*). A further advantage of the enclosed system is that noise is significantly reduced.

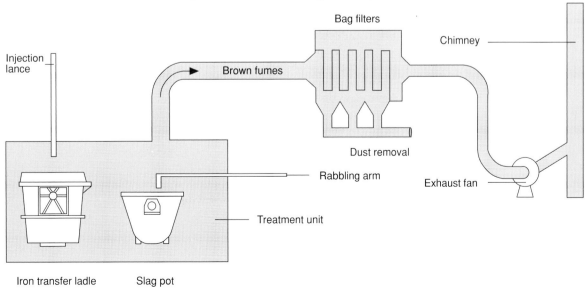

Figure 11 Transfer ladle desulfurisation plant

RS•C

As well as the environmental improvement, the new facility brought about a number of operational advantages.

▼   Installation of automatic systems for measuring the temperature of the iron, sampling it and storing the carbide reagent;

▼   the process is computer controlled;

▼   an increase in capacity of the works' desulfurising facility;

▼   increased efficiency in the use of the torpedo ladles as these are now filled to capacity rather than requiring a headspace to allow for the frothing associated with the injection process; and

▼   no build up of slag in the mouths of the torpedo ladles.

The last two points mean that the torpedoes can now carry 300 tonnes rather than 220 tonnes under the previous system.

These improvements have brought about savings which mean that the new facility will pay for itself in 4.6 years.

The design of the bag filtration system is shown in *Fig 12 (a)*.

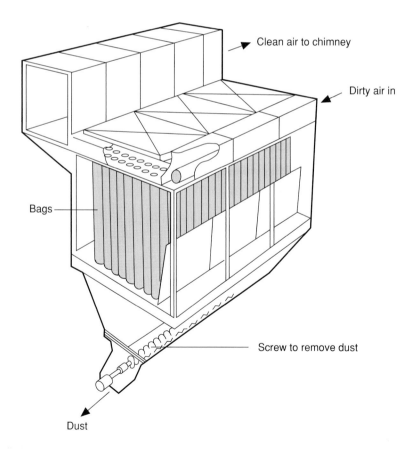

Figure 12 (a) A bag filtration plant

RS•C

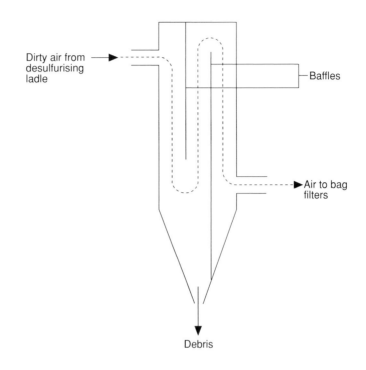

Figure 12 (b) A spark arrester

To chimney

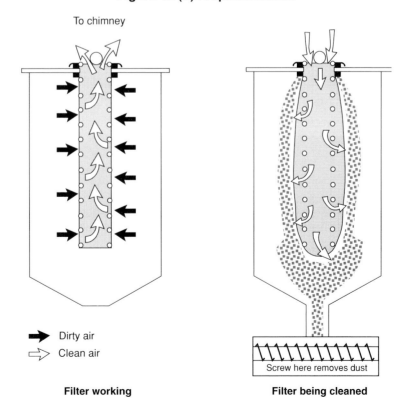

Filter working

Filter being cleaned

Figure 12 (c) Bag filters

RS•C

Each bag looks like a long sock made of polyester felt. Bags hang inside the filtration plant in eight groups of 200, a total of 1600 in all. The dirty air is sucked first through a spark arrester (*Fig 12 (b)*) containing baffles which prevent hot molten metal droplets and glowing particles getting through to the filters which could be damaged or even set on fire. Air is then drawn upwards through the socks, the outsides of which soon become coated with a cake of dust. It is this which actually does most of the filtering. The pressure drop across each filter is monitored and when it becomes too large, this indicates that the filter requires cleaning. Cleaning is carried out by blowing a burst of high pressure air through the filter in the reverse direction for a few milliseconds. As a result the accumulated dust is blown into a hopper. This system is called a pulse jet system. The dust contains some unreacted calcium carbide. This reacts with water to form flammable ethyne (acetylene) so, before disposal at a licensed landfill site, the dust is sprayed with water to ensure that this reaction has gone to completion.

Question 13
What is the equation for the reaction of calcium carbide (CaC_2) with water?

At the time of installation, the emissions from the filter plant were required by Her Majesty's Inspectorate of Pollution (HMIP), now The Environment Agency, not to exceed 100 mg of particulate matter per m^3 of air (adjusted to ambient temperature).

Question 14
The air emerges from the filtration plant at 30 °C. What would be the volume of 1 m^3 of the emerging air at ambient temperature (20°C)? Assume ideal gas behaviour and constant pressure.

However, it was anticipated that future legislation would specify 50 mg m^{-3} so the plant was actually designed to a limit of 30 mg m^{-3} to allow some leeway. In practice, emissions are closer to 10 mg m^{-3} and the scheme won a CBI Business Commitment to the Environment Award in 1992. Emissions are monitored by a system which measures absorption of light by the air. This is calibrated by a probe which sucks in a measured volume of air and weighs the particulates contained in it.

The terne-coated steel plant at Cookley

Terne-coated steel (the name comes from old French) is steel strip coated with nickel and a lead-tin alloy to prevent corrosion (*Fig 13*). It is used to make petrol tanks for vehicles and also gas meters.

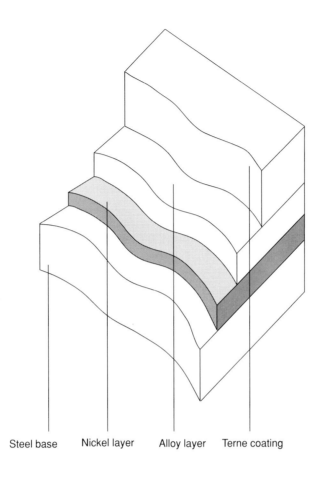

Steel base   Nickel layer   Alloy layer   Terne coating

Figure 13 Terne steel

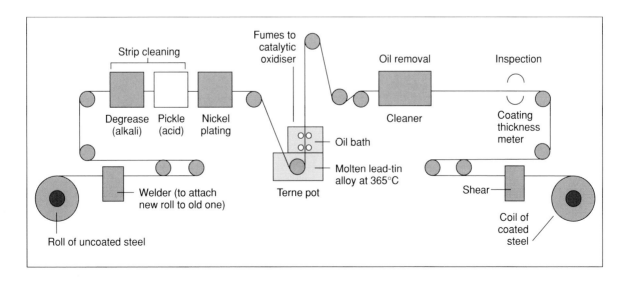

Figure 14 Schematic diagram of Cookley's terne coating line

RS•C

The coating process is shown schematically in *Fig 14*. A continuous roll of sheet steel is first degreased and pickled (cleaned in acid) and then electroplated with a thin layer of nickel which helps the lead-tin alloy to form a uniform coat. The strip then passes through a bath containing the molten lead-tin alloy at 365 °C. Emerging from this, it passes through rollers into a bath of synthetic oil. The rollers control the thickness of the coating and the oil assists adhesion and resists oxidation of the hot alloy. Residual oil is then removed with brush scrubbers and an alkaline detergent solution. The coating thickness is then monitored. The process is continuous; as one coil of steel is used up, another is welded on.

Oil fumes are generated in the oil bath as the hot, coated steel comes into contact with the oil. Some oil is vaporised, some thermally decomposed and some oxidised. This results in a blue haze, rather like that above an overheated chip pan, which is both visible and smelly. Analysis of the haze shows that it contains aldehydes and fatty acids with chain lengths from six to 19 carbon atoms as well as vapour of the unchanged oil. The aldehydes and acids result from the partial oxidation of the oil. Originally this haze was removed with a fume hood and discharged via a chimney. However, there were complaints from local residents about the smell and, from time to time, oil would condense and produce a fall out of droplets on cars and washing in the residential area around the plant.

Question 15

Draw the structural formula of one of the aldehydes and one of the fatty acids found in the haze. Which is the more oxidised?

Various options were tried over a number of years to treat the discharged air:

▼ electrostatic precipitation (Box 7) reduced the problem by an estimated 30 – 40%;

▼ perfume was sprayed into the extraction system;

▼ knitmesh filters of steel wool were used; and

▼ water scrubbing (Box 8) produced an estimated 80% improvement but resulted in an oil-water emulsion which could not be separated and had to be disposed of to landfill.

RS•C

Electrostatic precipitation   Box 7

Here, the dirty gas passes through a large box containing high voltage discharge wires and charged collecting plates – see below. The oil droplets (or other particles) pick up an electric charge and are attracted to the plate with the opposite charge. This is rather like the way in which small pieces of paper are attracted to a plastic ruler which has been charged by rubbing it on dry, nylon clothing.

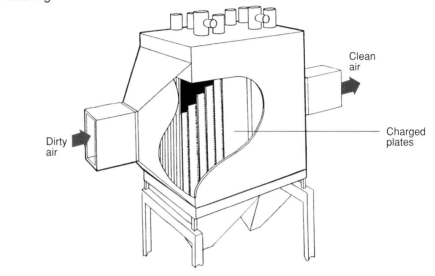

Water scrubbing   Box 8

Here the dirty air is sprayed with high pressure water jets which clean it. The air-water mixture is thrown against the side of a cylindrical tank and moves up the inside of the tank in a spiral – see below. The dirty water runs down the side of the tank and clean air escapes from the top. This is called a cyclone and the principle is rather like that of a spin drier. A potential problem is that this can exchange one type of pollution (air) for another (water).

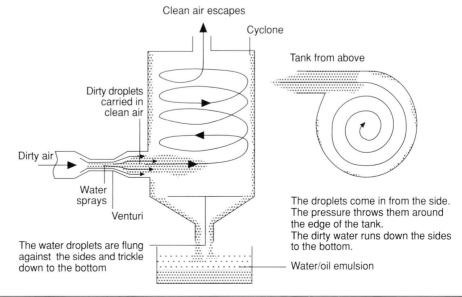

None of these was wholly satisfactory. A high temperature incinerator was considered in which the fumes would be mixed with more air and passed through a gas flame at 760 °C to completely oxidise them to carbon dioxide and water. However, this would have been relatively expensive to run because of fuel costs. Eventually, in 1991, a catalytic oxidiser was fitted. Here the fumes were burned in a gas flame at 370 °C and then passed through a fluidised bed of aluminium oxide pellets impregnated with a chromium oxide-based catalyst system to complete the oxidation at low temperature (*Fig 15*). The catalyst is not sensitive to poisoning by metals such as iron and lead, although a small amount of catalyst is lost as dust blown out with the discharged gas.

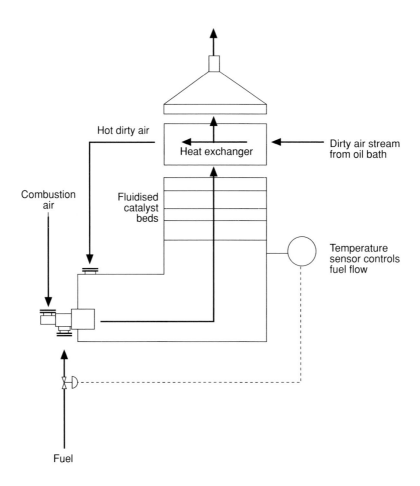

Figure 15 (a) The catalytic oxidiser

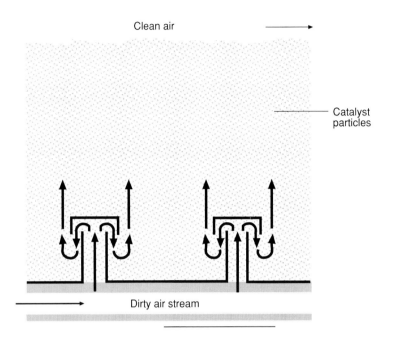

Clean air

Catalyst particles

Dirty air stream

Figure 15 (b) The fluidised bed

The catalyst saves fuel costs by operating at a lower temperature than a conventional incinerator. Further fuel savings are made by passing the incoming dirty air through a heat exchanger so that it is warmed up by the hot, outgoing, clean air. When the system was installed, there was some experimentation to find the optimum temperature and flow rate to minimise both pollution and fuel use. It is now considered that this system reduces emissions by 98–99%. Complaints are now rare except in weather conditions which cause a temperature inversion which traps a layer of air close to the ground. The limit for discharge of volatile organic compounds (VOCs) agreed with the Local Authority is 50 mg of carbon per m^3 of air and typically the plant discharges 20–25 mg of carbon per m^3. This figure refers to the sum of all organic compounds taken together and is measured by a flame ionisation detector such as is used in many gas chromatographs.

There is now, however, concern about discharge of carbon monoxide caused by incomplete oxidation. This is a different problem to the original one, which concerned pollution that local residents could smell and see. Carbon monoxide is, of course, colourless and odourless. One possible solution would be to add an extra, different, catalyst system to oxidise the carbon monoxide to carbon dioxide but it is possible that the new catalyst might be poisoned by dust from the original catalyst system.

Another, more fundamental, solution to the problem is therefore under consideration. This is to eliminate the oil-lubricated rollers which control the coating thickness and replace them with jets of high pressure air called air knives which would scrape off excess lead-tin coating. These require no oil and therefore eliminate the problem. This system would entail significant capital cost and lost production as the new plant was installed and run in. Tests suggest that the air knife system leaves some porosity in the coating which would therefore need to be sealed. It is quicker than the present system and seven days worth of production could be achieved in five.

RS•C

Answers to questions

1. $Fe_3O_4(s) + 4CO(g) \rightarrow 3Fe(l) + 4CO_2(g)$

     $Fe_3O_4(s) + 4C(s) \rightarrow 3Fe(l) + 4CO(g)$

2. $Fe^{III}_2O^{-II}_3(s) + 3C^{II}O^{-II}(g) \rightarrow 2Fe^0(l) + 3C^{IV}O^{-II}_2(g)$

     Iron is reduced from III to 0 and carbon is oxidised from II to IV.

     $C^{IV}O^{-II}_2(g) + C^0(s) \rightarrow 2C^{II}O^{-II}(g)$

     This is a reverse disproportionation reaction in which carbon is oxidised from 0 to II and reduced from IV to II.

     $Fe^{III}_2O^{-II}_3(s) + 3C^0(s) \rightarrow 2Fe^0(l) + 3C^{II}O^{-II}(g)$

     Iron is reduced from III to 0 and carbon is oxidised from 0 to II.

     $C^0(s) + O^0_2(g) \rightarrow C^{IV}O^{-II}_2(g)$

     Carbon is oxidised from 0 to IV and oxygen is reduced from 0 to -II.

3. $CO(g) + \frac{1}{2}O_2(g) \rightarrow CO_2(g)$

     $\Delta H^\ominus = -283.0$ kJ mol^{-1}.

4.   Sensible suggestions would include:

     MgS, CaS and CaSO$_4$ Na$_2$S and Na$_2$SO$_4$, CaS and CaSO$_4$. Bear in mind that the conditions for the reactions are far from room conditions.

5. $CaCO_3(s) \rightarrow CaO(s) + CO_2(g)$

     $\Delta H^\ominus_{reaction}$ $= -393.5 -635.1 +1206.9 = +178.3$ kJ mol^{-1}

     $\Delta S^\ominus s_{ystem}$ $= 213.6 + 39.7 -92.9$ J K^{-1} mol^{-1} $= 160.4$ J K^{-1} mol^{-1}

     $\Delta G^\ominus$ $= 178.3 - \dfrac{(1500+273) \times 160.4}{1000}$

                        $= -106.1$ kJ mol^{-1}

     So the reaction is feasible.

     The assumption made is that the vaues of $\Delta H^\ominus$ and $\Delta S^\ominus$ are the same at 1500 °C as they are at standard conditions.

6.   The gas would be reduced to 32.8% of its original volume.

7. $4NH_3(g) + 3O_2(g) \rightarrow 2N_2(g) + 6H_2O(g)$

8.   24 025.4 kJ

9.   Carbon monoxide and hydrogen.

10.   a) $CaS(s) + 2H_2O(l) \rightarrow H_2S(g) + Ca(OH)_2(aq)$

      b)   The water is acting as an acid.

      c)   Adding lime (a base) would therefore suppress the reaction.

11.   a) $SO_2(g) + H_2O(l) + \frac{1}{2}O_2(g) \rightarrow H_2SO_4(aq)$

      b)   Oxygen and water are required.

12.   Droplets would have a greater surface area exposed to the air than undisturbed molten iron.

RS•C

13. $CaC_2(s) + 2H_2O(l) \rightarrow HC{\equiv}CH(g) + Ca(OH)_2(aq)$

14. $0.97 \ m^3$.

15. Any aldehyde (RCHO) and carboxylic acid (RCO_2H) from C_6 to C_{19}. The acid is the more oxidised.

RS•C

The story of BRL 46470A

Bringing a new medicine to market takes a great deal of time and money. Typically it can cost £150 million and take up to 16 years. Only one compound in 10 000 actually makes it! *Figure 1* illustrates the process. This case study uses the substance code-named BRL 46470A, a potential drug substance discovered in 1986 by SmithKline Beecham, as an example to show some of the steps involved in the development of what the pharmaceutical industry calls a new chemical entity (NCE). The study looks particularly at different methods of synthesis.

▼ The method by which it was initially synthesised in the laboratory on a scale of a few grams for initial screening – **the discovery route**.

▼ The method developed for producing it on a scale of a few kilograms for full clinical trials – **the supply route**.

▼ The method developed for manufacturing it for eventual marketing on a scale of 10–100 tonnes per year in a factory – **the manufacturing route.**

Medicines and drugs   **Box 1**

There is often confusion about these two terms, and the word drug is now often associated with illegal substances such as ecstasy, cocaine and heroin.

Strictly speaking, a drug is a substance which affects how the body works – either for better or worse. A medicine is a substance that improves health. Medicines contain beneficial drugs (which are the active ingredients) as well as other substances which make them easy and convenient to take.

A newly synthesised chemical which may be a drug and may eventually form the active ingredient of a medicine is often called a new chemical entity (NCE).

[BRL 46470A therefore began life as an NCE, was shown by testing to be a drug and would have become part of a medicine had it ever been marketed.]

RS•C

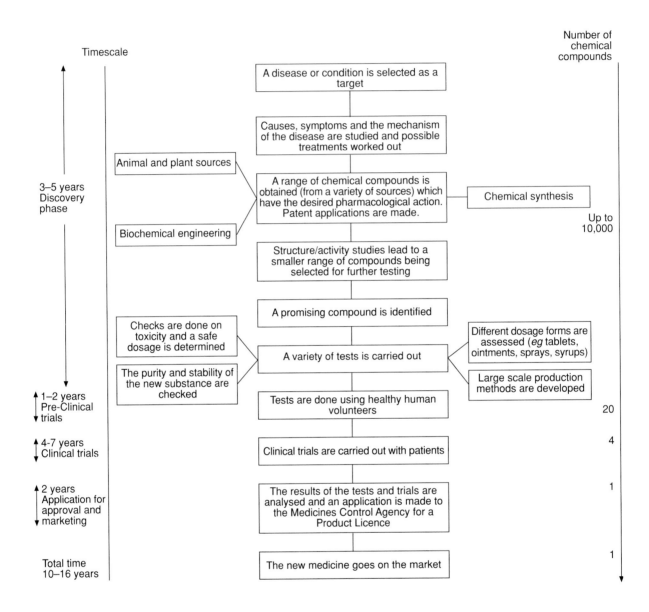

Figure 1 Stages in the development of a medicine
(modified from *Discovery and development of a new medicine*, The Association of the British Pharmaceutical Industry)

In fact BRL 46470A was not marketed as a decision was taken to terminate its development in August 1993. It was, to use the jargon, prioritised out. Prioritisation of portfolios is undertaken on a regular basis. It involves assessing the commercial promise of various compounds in the light of experimental results and the products of other companies.

When a drug is prioritised out, work on it does not necessarily stop immediately. Work in progress will be tidied up to maximise the value of the drug (and associated data) as an asset should an opportunity for out-licensing (manufacturing by another company) arrive.

RS•C

In the beginning

BRL 46470A entered development as an anti-emetic – a medicine which reduces the vomiting that, for example, often accompanies chemotherapy (drug treatment) of cancer. It was already known that chemicals which bind to so-called 5-HT$_3$ receptors in cells and disrupt their normal function (called 5-HT$_3$ receptor antagonists (Box 2) have an anti-emetic effect, and the likely structure of molecules that do this was already known. Work on 5-HT$_3$ antagonist drugs was already well advanced – eg Granisetron (SmithKline Beecham) and Ondansetron (GlaxoWellcome) were in the late stages of clinical trials. BRL 46470A was developed as a back up compound to Granisetron (BRL 43694). Back-ups are substances similar to compounds of known activity. They are developed as an insurance against some unforeseen problem with the first candidate. The similarities in structure between BRL 46470A, Granisetron and Ondansetron can be seen in Fig 2.

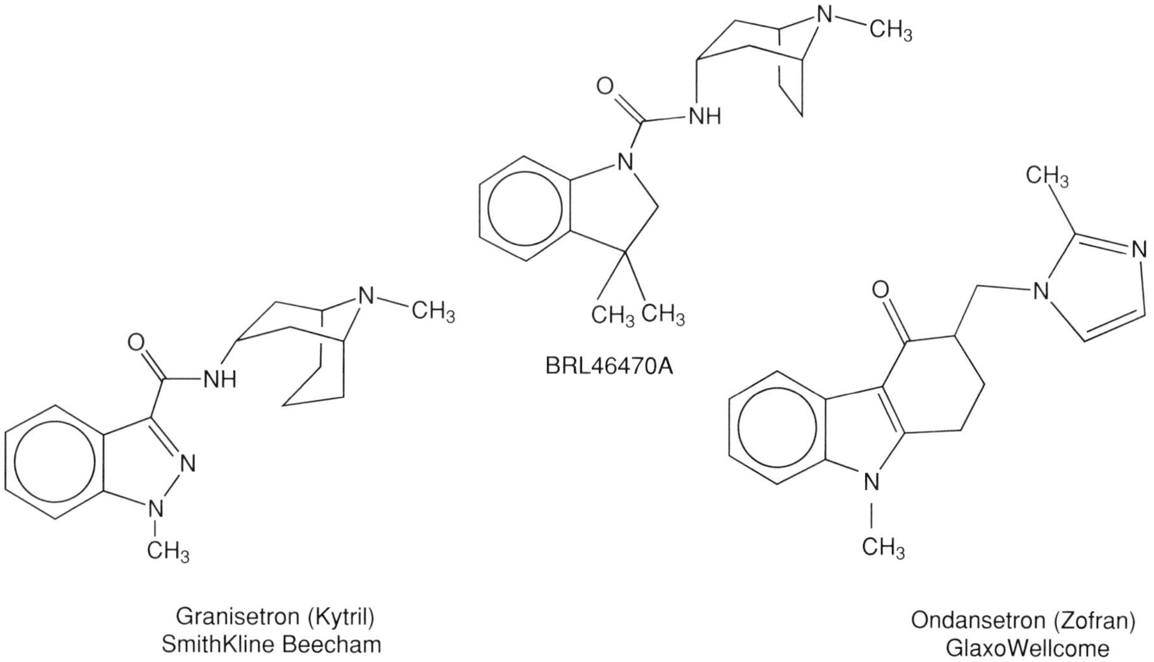

BRL46470A

Granisetron (Kytril)
SmithKline Beecham

Ondansetron (Zofran)
GlaxoWellcome

Figure 2 Strucures of BRL46470A, Granisetron and Ondansetron.
All three drugs are supplied as hydrochlorides (in which the tertiary amine is
protonated) to make them water soluble

RS•C

5-HT$_3$ receptors   **Box 2**

We believe that drug molecules interact with specific reactive sites – molecules or parts of molecules within the body. These sites are called receptors and are typically large biological molecules (often proteins) located in the outer membranes of cells. They control the flow of chemicals into and out of the cell. **Agonists** are molecules which bind to receptors and produce a biological response. **Antagonists** bind to the receptor and produce no response but block the receptor to the agonist. 5-HT$_3$ receptors are ion channels which conduct Na^+ and K^+ ions through the cell membrane. They are found in the right ventricle of the heart, the gastrointestinal tract and the central nervous system (CNS). They normally bond to the molecule serotonin (5-hydroxytryptamine, 5HT, see below)). 5-HT$_3$ antagonists have been shown to work as anti-emetics (medicines which prevent vomiting). Because these receptors are located in the CNS, their antagonists might also be expected to act as anti-psychotic, anxiolytic (anxiety reducing) and cognitive-enhancing drugs.

Serotonin

Question 1

Look at the structures of BRL 46470A, Granisetron and Ondansetron in *Fig 2*. What structural features do they have in common?

   Compare the stuctures with that of serotonin.

   Once a compound such as BRL 46470A has been synthesised and shown to have some pharmacological activity (*ie* an effect on the body), it is routine practice to make a number of structurally related compounds, test their activity and develop a **structure-activity relationship**. This is an attempt to maximise the activity of the compound, tailor its pharmacological activity (for example by minimising interaction with other receptors) and gain an understanding of the nature of the receptor. For example, it might be the case that substituting an electronegative atom such as a halogen in place of a hydrogen has a particular effect which becomes more pronounced with greater electronegativity, or that substituting a bulkier group for a hydrogen might have a different effect related to the size of the substituent. The structure-activity relationship for BRL 46470A is described in Box 3.

The structure-activity relationship   **Box 3**

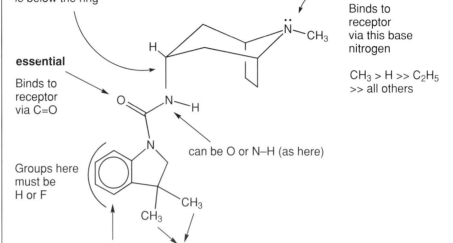

This amine must be axial
ie below the ring

essential

Binds to
receptor
via this base
nitrogen

$CH_3 > H >> C_2H_5$
$>>$ all others

essential

Binds to
receptor
via C=O

can be O or N–H (as here)

Groups here
must be
H or F

lipophilic   lipophilic

NB Lipophilic means oil liking – *ie* having an
affinity for non-aqeous solvents, rather than water

BRL 46470A Structure activity relationship

This is an attempt to relate the structure of a compound to its pharmacological
activity.

BRL 46470A is a 5-HT$_3$ receptor antagonist – that is it binds to the receptor but
does not activate it. The binding is believed to take place via the basic nitrogen
atom (top right of the structural formula as drawn above) and the carbonyl
group. The aromatic ring is also involved but in a way which is not fully
understood. Some of the main features of the structure-activity relationship are
marked on the structure.

Question 2

Look at the structure of BRL 46470A in Box 3. Explain in terms of the number and
arrangement of electrons, how an oxygen atom (O) can replace an N-H group.

At this stage, two things are important.

▼   Many related compounds have to be synthesised in small quantities (around
500 mg or less) for investigation. So a *general* synthetic route which can be
used to prepare a number of similar compounds is preferred (Box 4).

▼   The compounds need to be screened for activity relatively quickly and
cheaply. For example, it is possible to measure binding to the 5-HT$_3$ receptor
in vitro (literally in glass – *ie* in a test-tube experiment), rather than in a trial on
living organisms – *in vivo*.

RS·C

General and specific methods of synthesis   **Box 4**

One of the intermediates in the synthesis of BRL 46470A is 3,3-dimethylindoline.

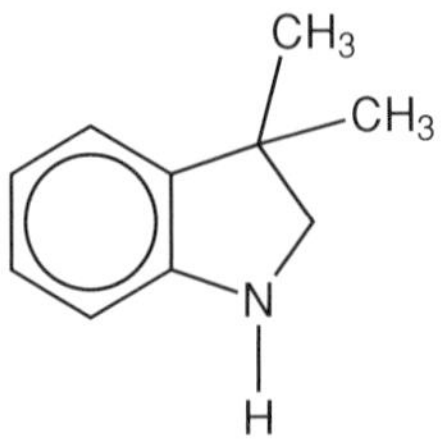

3,3-Dimethylindoline

Two possible methods (A and B) of synthesis of 3,3-di-substituted indolines are shown. Method A is the more flexible as a range of 3,3-di-substituted indolines with different R' groups can be prepared by a suitable choice of R'.

Method B is a poor *general* method. This is because, even if a suitable starting material can be found with the appropriate R groups, mixtures of products are possible (unless X=H). This is because there are two possible modes of ring closure as shown by the dotted lines.

For BRL 46470, X=H and a suitable starting material is available. This made method B a suitable *specific* method for the synthesis of 3,3-dimethylindoline.

Method A was used as part of the discovery route to BRL 46470A while method B was used as part of the supply route and conditions were devised which produced the desired ring closed product in 99% yield.

Method A

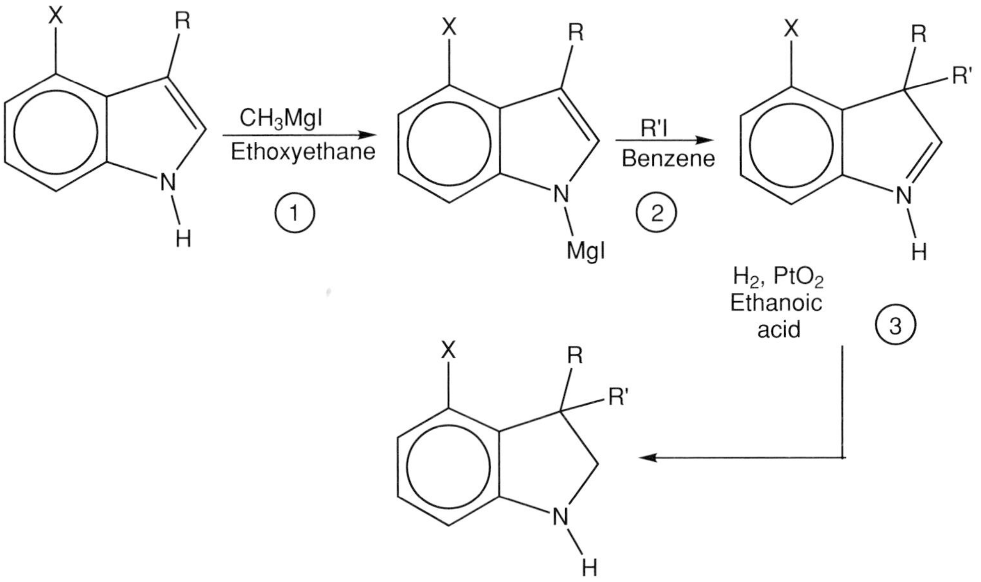

3,3-Disubstituted indoline
(if X=H)

Box 4 continued

Method B

either

or

The dotted lines show
two possible methods
of ring closure giving
two possible products
(unless X=H)

This can be converted
to the indoline by
reduction of the C=0 group

The next step

Once a single compound such as BRL 46470A has been selected for progression as a potential medicine, it has to successfully overcome a large number of commercial and technical hurdles. Commercial considerations include the following.

▼   What is the size of the global market for this type of medicine?

▼   What is the likely market share that this compound could attain?

▼   Will the product be first to market, *ie* be on sale before other competitive products?

▼   What advantages will the new product have over the likely competition?

▼   What is the compound's patent position, *ie* how many years of patent are left? Medicines whose patent has expired are called generic medicines and can be made and sold by anyone.

▼   What is the marketing expertise of the company for this type of medicine?

Technical considerations include.

▼   Does it have the desired **clinical effect**?

▼   Is it **toxic?** This includes acute toxicity (is it immediately poisonous), long term (chronic) toxicity, carcinogenicity (could it cause cancer?), teratogenicity (could it cause birth defects?) *etc.*

▼   Is it **safe** for the intended market?

▼   Is it **bioavailable** (does a significant percentage of the dose reach the target organ intact)?

▼   What is the best **route of administration** – orally, intravenously *etc*?

RS•C

▼   Is the **pharmacokinetic profile** (for example the time the drug remains in the body) suitable for the intended dosing regime – how often it needs to be taken?

▼   What is the **fate** of the drug after it has been taken (how is it metabolised in the body)?

▼   Can the drug be **prepared** safely and easily?

▼   How will it be **formulated** (tablets, capsules, patches, injected, as a syrup *etc*)?

Once a decision has been made to progress with the development of a drug, it will be subjected to a battery of trials to assess its toxicology, pharmacokinetics and clinical efficacy. Toxicological trials test whether the drug is poisonous in the long and short terms. Pharmacokinetic trials measure what happens to the drug in the body. For example:

▼   is it stable in the gut (which contains hydrochloric acid and digestive enzymes)? If not, the medicine will not be able to be administered orally – *ie* by mouth; and

▼   what happens to the drug in the bloodstream? How long does it remain there and how is it broken down and excreted? What are the breakdown products and what effects do they have?

Drugs which are unstable in the gut and cannot be taken by mouth may have to be injected directly into the blood stream, by-passing the gut. This means that they will have much more limited usefulness; for example they will not have the potential to be over-the-counter medicines. An example of this type of medicine is insulin. It has to be injected, which means that it has to be used with medical supervision or that the patient has to be specially trained to administer it to him or herself. A medicine for migraine, for example, would need to be able to be taken orally as patients are rarely hospitalised for this condition. However, injectable-only medicines are acceptable for acute conditions such as vomiting during chemotherapy where a patient is likely to be in hospital.

The timescale over which a medicine takes effect is also important. A medicine to treat acute pain must work within a few minutes at most, whereas a medicine to treat a chronic condition such as depression may be acceptable even if it takes several days to produce an effect.

Compounds with one type of activity may sometimes have more than one clinical application. For example, the drug imipramine is used as an antidepressant but is also used to treat bed-wetting in children. Such a crossover occurred to BRL 46470A. As well as its anti-emetic effect it was also found to show activity in a model for anxiety – and would therefore be expected to have an anxiolytic (anxiety-reducing) effect. This was not unexpected given the distribution of the 5-HT$_3$ receptors, on which BRL 46470A acts, in the central nervous system (Box 2). In 1989, BRL 46470A was selected for progression as an anxiolytic.

The development plan

At this stage, a number of activities have to be completed satisfactorily if the medicine is to reach the market place. These include:

▼   toxicology trials;

▼   pharmacokinetic studies;

▼   clinical trials;

▼   determining metabolism;

▼   filing patents (Box 5);

▼   filing applications to government regulatory authorities;

▼   market research and strategy planning;

▼   developing suitable formulations; and

▼   developing a suitable route of manufacture.

Patenting   **Box 5**

A patent gives a company or an individual the right to exclude others from commercialising an invention. Patents normally last 20 years, after which anyone can make and sell the medicine. So the inventing company needs to recoup its development costs and make a reasonable profit within this time, as afterwards other firms can make and sell the medicine without having had to pay for its development. Typically, it may take 10 years of a 20 year patent life before the medicine is on the market, leaving just 10 years for the company to recoup its development costs and make a profit. Extension of a patent's term for a further five years is now available for pharmaceuticals in Europe, the US and Japan.

The length of a patent's life means that the decision as to when to patent a medicine can be a difficult one. Patenting early secures the company's right to the medicine but, as patents are in the public domain, it may give competitors an insight into your research strategy. It also means that much of the medicine's development will take place within the life of the patent, thus reducing the time available to sell the medicine exclusively. The later the date of patenting, the longer the exclusive selling time. However, patenting too late can be disastrous; if another company happens to have been working on the same idea and patents it just one day before you do, all the development costs may be wasted.

In general, because the main areas of pharmaceutical research are well known, and competition is very fierce, early patenting is highly advisable.

A plan must be devised to carry out all the above operations in parallel so that the medicine can be brought to market as quickly as possible. It is important to remember that at this stage the development of the medicine is costing the company a considerable amount of money, none of which will be recouped unless and until it is sucessfully commercialised. Typically it could cost £150 million (at 1996 prices) and take 10 years to bring a medicine from discovery to market (*Fig 1*). The plan must be subjected to critical path analysis to eliminate bottlenecks which might hold up other parts of the study. For example clinical trials, which involve many doctors, and patients or volunteers are the most expensive, so it is important that these are not delayed. It is therefore vital that sufficient supplies of the drug are made available for clinical trials – at this stage, availability of the compound is more important than its cost.

Many compounds fall by the wayside during the various stages of development as shown by *Fig 3* which indicates the number of compounds remaining in development after each stage (the attrition rate).

RS•C

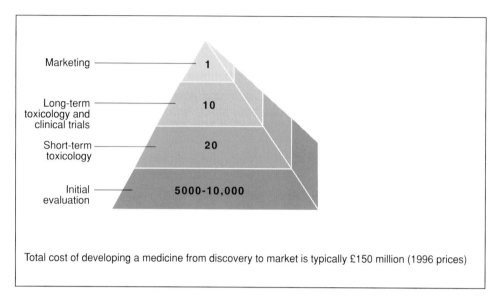

Figure 3 Attrition rate of compounds

Discovery, supply and manufacturing routes

The route by which a compound like BRL 46470A is synthesised is likely to change as it proceeds through development. The first supplies of drug for trials will probably be made by the discovery route to save time. As time goes by, the responsible chemists will hone this route making it more efficient and therefore cheaper. For example, they might find that it is possible to use less of one or more reagents, they might find that a cheaper solvent can be used, a shorter time of heating to a lower temperature might be found to be feasible *etc*. However, the potential for such improvements is limited and it is normally found that costs approach a minimum (*Fig 4*). So while initial supplies for trials are being made by the discovery route, development chemists will be trying to devise a more efficient route which will be intrinsically cheaper and can be used on a larger scale – the supply route.

RS•C

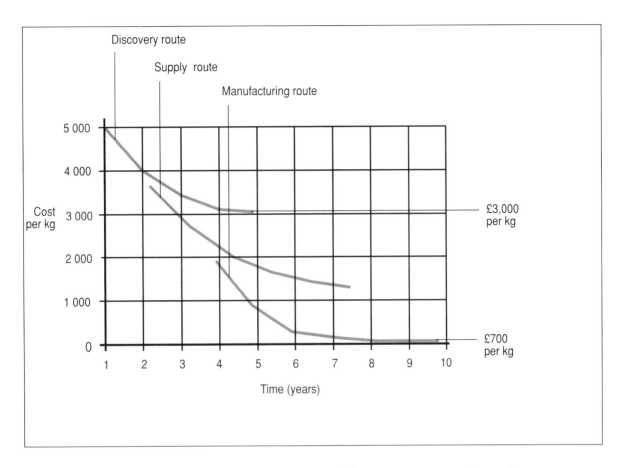

Figure 4 Supply and discovery routes cost graph

Unlike the discovery route, the supply route can be specific for the one compound of interest. The chemists responsible for developing the synthetic methods will do a literature search to discover ways of making the required compound or related ones. They will then try out one or more methods. Factors they must consider include:

▼ the yield of each stage;

▼ the potential for scaling up as the basis for the manufacturing route;

▼ the method of isolating the required product from solvents *etc*; and

▼ the method of purifying the product.

Different considerations apply to isolating the product in a pilot or manufacturing plant compared with the laboratory. For example, in a plant, crystallisation of a product followed by filtration is much preferred to evaporating off a solvent on a rotary evaporator. The latter method is likely to leave the product spread thinly over the inner surface of the vessel making it difficult to get out. Scraping with a spatula, as one might use in the laboratory, is not practical with plant-scale apparatus!

Purity of the drug is an interesting issue. Most drugs will be greater than 98% pure, (although some antibiotics may be as low as 80%). Impurities include solvents, water and minor reaction by-products. A new method of synthesis may well result in different impurities which may require toxicology tests to be repeated. Each route is said to produce a different impurity profile.

RS•C

Scaling up

Even if the discovery route were to be used in a pilot plant, scaling up a reaction is more complex than simply multiplying all the quantities by an appropriate factor (Box 6).

The mathematics of scaling   Box 6

Increasing the volume of a vessel by a factor of 1000 means scaling up its linear dimensions by a factor of 10 and this means that its surface area will be increased by a factor of 100. This in turn will affect the rate at which heat can be lost from the vessel, as heat loss occurs from the surface and will be proportional to the surface area.

Reaction vessel of volume 1 dm^3 (1000 cm^3)

Volume = 10 cm x 10 cm x 10 cm = 1000 cm^3 = 1 dm^3

Surface area = 6 x (10 cm x 10 cm) = 600 cm^2

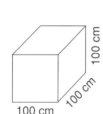

Reaction vessel of volume 1000 dm^3

Volume = 100 cm x 100 cm x 100 cm = 1 000 000 cm^3
$$= 1000 \ dm^3$$

Surface area = 6 x (100 cm x 100 cm) = 60 000 cm^2

The volume increased by a factor of 1000 while the surface area increased by a factor of 100

So increasing the volume of a reaction vessel by a factor of 1000 will increase the quantities of reactants by a factor of 1000 and, if the reaction is exothermic, the heat generated will also increase by a factor of 1000. However, the surface area will increase only by a factor of 100 making heat loss more difficult, and possibly leading to a dangerous increase in temperature.

Another example is stirring. Different types of stirrer vary widely in their ability to mix. What runs well in the laboratory with, say, a magnetic stirrer may not work with a paddle stirrer in a large reaction vessel.

In general, when scaling up a synthetic route, it is necessary to run it in the plant on a trial basis to validate the process.

While devising a supply route, development chemists are thinking ahead to devising a manufacturing route. There are several further considerations here.

▼   How much drug material will be required? (Box 7).

▼   Limitations of manufacturing plant, for example size of vessels, maximum and minimum temperatures available.

▼   Cost and availability of raw materials.

RS•C

- ▼ Capital investment. The drug might be made in a general pupose plant capable of being used to make a variety of compounds or it might require a purpose-built one.

- ▼ The availability of special facilities, for example to carry out hydrogenation or fermentation.

- ▼ The energy balance of the process – which includes energy required for heating reactions and heat loss from exothermic ones (Box 8).

- ▼ Safety. Large scale manufacturing involves greater risks than laboratory procedures because of the larger quantities of material involved and because the operatives are not such skilled chemists. For example, a solvent such as ethoxyethane (diethyl ether, ether) cannot be used on a plant because of its volatility and low flash point whereas it can be used in the laboratory if appropriate precautions are taken.

- ▼ Environmental issues; for example disposal of by-products, the type of waste (solid or liquid) and the method of disposal (incineration, landfill *etc* (Box 9).

- ▼ The reproducibility and robustness of a synthesis (Box 10).

How much drug is required   **Box 7**

This depends on the the number of people who suffer from the condition treated by the medicine, the percentage of these who are treated using *this* medicine and the dose of medicine. The number treated in turn depends on cost, how much more effective the medicine is than its competitors, and the skill with which the drug is marketed. The required quantity might vary between 20–30 kg per year to 200–300 tonnes. Typical single dosages for drugs for humans vary from tens of micrograms to hundreds of milligrams.

Energy balance   **Box 8**

All reactions exchange energy, usually in the form of heat, with their surroundings. In the laboratory this is not normally a problem. We can supply the heat needed to make an endothermic reaction go (or to speed up a slow reaction) with a heating mantle. We can cool an exothermic reaction under the tap or by using an ice bath. Heat exchange poses more of a problem in larger vessels. They lose heat more slowly than small ones due to their smaller surface area to volume ratio (Box 6). This means that an exothermic reaction may be generating heat faster than it can be lost, leading to a rise in temperature. This speeds up the reaction which then gives out heat even faster, causing a potentially dangerous runaway.

Pilot plants may have a reservoir of coolant liquid held at a low temperature which can be used to quench such a reaction. Also it is important to understand the thermodynamics of reactions before scaling them up, and one of the important jobs for the development chemist is to do careful calorimetry experiments on all reactions before they are scaled up.

On a plant, the energy needed to heat up a reaction may be a significant part of the cost of the process and needs to be kept to a minimum.

RS•C

Landfill tax **Box 9**

On 1st October 1996 the UK government introduced a landfill tax. This is imposed on most categories of rubbish and is aimed to encourage industry to:

produce less waste, recover more value from waste (for example through recycling or composting) and to use more environmentally friendly methods of waste disposal.

The tax is levied at £2 per tonne for inert waste (non-polluting waste which will not decay to produce methane) and £7 per tonne for other waste. The tax will raise £450 million for the government. This increased cost of landfill has made many companies rethink their methods of waste disposal.

Robustness and reproducibility of syntheses **Box 10**

Some reactions are more sensitive than others to small changes in conditions such as quantities of reagents, temperature, solvents *etc.* Such changes may affect the yield and purity of the required product by encouraging the formation of side products. Ideally, on a plant, one reqires a synthetic method which is insensitive to such small changes and always gives the same result so that the quality of the product is not affected. Such a synthetic method is referred to as robust and reproducible.

Designing the synthesis

Designing the manufacturing method can take a team of 10 synthetic and analytical chemists up to five years during which time the drug is being produced for testing by the supply route.

We will look briefly at some of the synthetic methods used for BRL46470A.

Chemical intuition suggested that the best overall synthetic strategy was to couple together the compounds 3,3-dimethylindoline and tropane amine via a carbonyl group (*Fig 5*). So methods were required to make these two precursors.

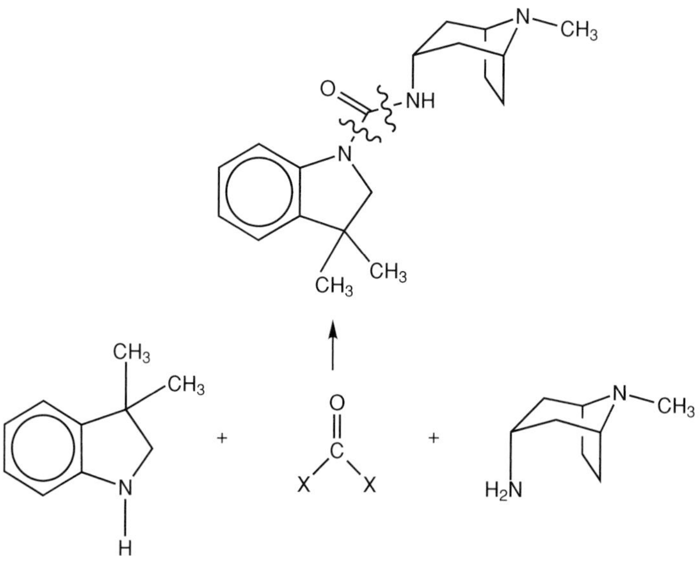

3,3-Dimethylindoline   Tropane amine

Figure 5 The BRL 46470A synthetic strategy

Synthesis of 3,3-dimethylindoline

The discovery route

The discovery route to this compound is shown in *Fig 6*. The first step is reaction with a Grignard reagent, methyl magnesium iodide, carried out in ethoxyethane solution. This is followed by treatment with iodomethane in benzene, with reduction by hydrogen in ethanoic acid solution as the final step. None of these steps is suitable for scaling up – ethoxyethane is highly flammable and has a low flashpoint, benzene is a carcinogen and hydrogen is explosive and therefore requires special handling. Ethanoic acid is a difficult solvent to remove because of its relatively high boiling temperature (391 K). The alternative of crystallising the 3,3-dimethylindoline requires the addition of sufficient base to neutralise the acidic solvent – also prohibitive. Moreover the overall yield (50% x 70% = 35%) is poor. In fact this route was never used to prepare BRL 46470A except during the discovery phase.

Question 3

How would you expect dimethylindoline to react with ethanoic acid? Use your answer to help you explain why it is neccessary to add a base before crystallising 3,3-dimethylindoline.

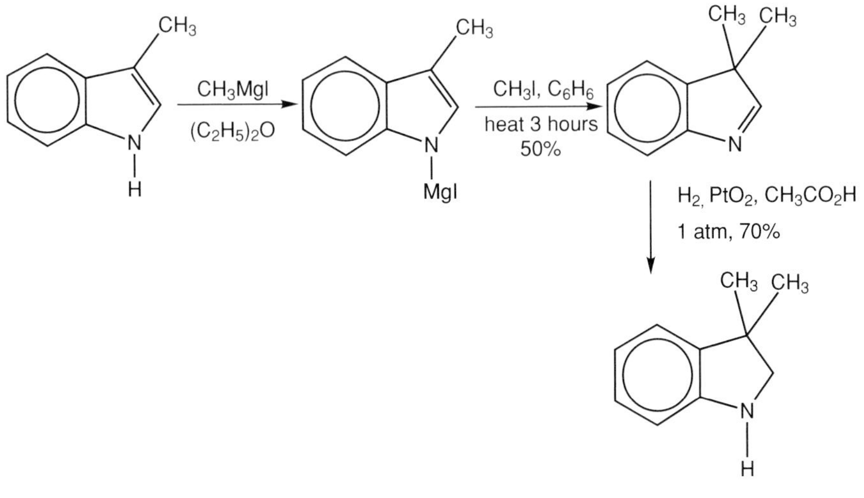

Figure 6 The discovery route to 3,3-dimethylindoline

RS•C

The supply route

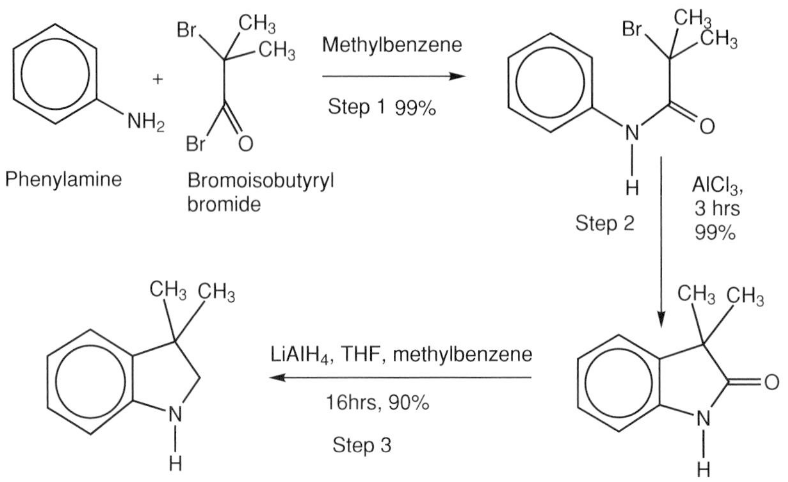

Figure 7 The supply route to 3,3-dimethylindoline

A search through the chemical literature revealed the method shown in *Fig 7* starting from phenylamine (aniline) which is cheap and readily available. Steps 1 and 2 were ideal, having extremely high yields and relatively innocuous reagents and solvents. Step 3 posed some problems, however. Although the yield was quite high, it used lithium tetrahydridoaluminate (III) (lithium aluminium hydride, $LiAlH_4$) which is water sensitive (giving off hydrogen when in contact with water) and it also took 16 hours of expensive plant time. Of most concern, however, was the fact that under certain conditions the reaction had a tendency to run away due to accumulation of unreacted starting material. This caused a sudden release of energy. Although careful study of the reaction led to successful methods of controlling it, the stringent controls required were thought to be unsuitable for manufacturing scale. Another slight disadvantage of this method was that one of the starting materials, bromoisobutyryl bromide, is a dibromo- compound. Although this is relatively cheap on a cost per kg basis ($10.64 per kg) it is considerably more expensive on a cost per mole basis due to its high relative molecular mass. Neither bromine atom is present in the final product so one is paying for two bromine atoms which go down the drain. A total of approximately 100 kg of BRL 46470A was prepared by this route for various uses – toxicology, formulation and clinincal trials.

Question 4
Lithium tetrahydridoaluminate (III) reacts by generating hydride ions (H–). Why does this reagent reduce the C=O group but not the aromatic ring? Give two reasons.

Question 5
Work out and compare the costs per mole of bromoisobutyryl bromide ($10.64 per kg) and of phenylamine ($2.45 per kg).

Question 6
What is the systematic name of bromoisobutyryl bromide (*Fig 7*)?

RS•C

The manufacturing route

1 | Methylbenzene reflux
2 | HCl (g)
3 | NaBH₄
4 | H₂O

One-pot
70–80%

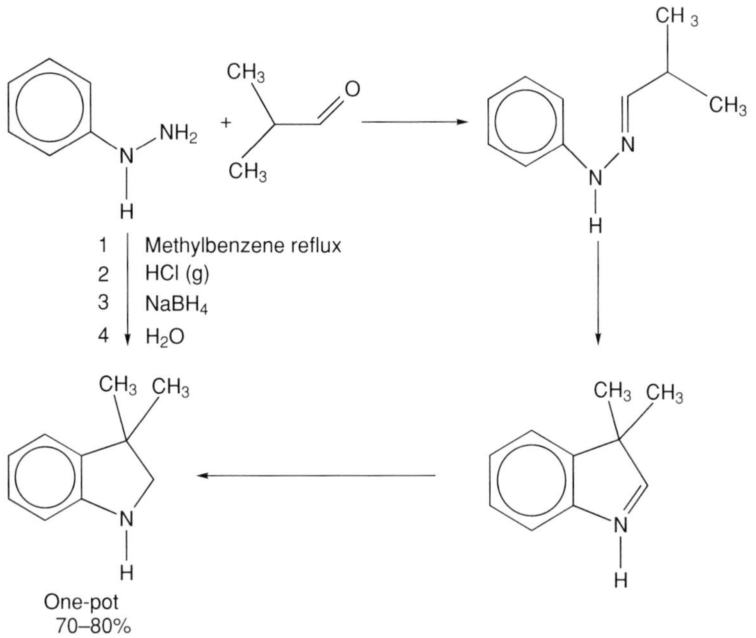

Figure 8 The manufacturing route to 3,3-dimethylindoline

As BRL 46470A has never made it to the manufacturing stage, no final decision was taken on the manufacturing route to 3,3-dimethylindoline. One possibility under consideration was that shown in *Fig 8*. This is a "one pot" synthesis, meaning that none of the intermediate products need to be isolated, giving considerable savings in cost. Notice also that the reagents and conditions are relatively innocuous; methylbenzene is less toxic than benzene, sodium tetrahydridoborate(II!) (sodium borohydride, NaBH $_4$) is a safer reducing agent than either hydrogen or lithium tetrahydridoaluminate(III) and only a small amount of cooling is required. So the method suffers from none of the inherent problems of the other two routes.

Similar processes and considerations would apply to the synthetic methods adopted for the other material required for the manufacture of BRL 46470A, tropane amine.

Costing the process

It is vital to be able to accurately know the cost of any industrial process and, in particular, to be able to predict the effect of changes in the process on the final cost of the product. This is done on a spreadsheet so that the effect of various changes can be assessed. For example, the effect of:

▼   change in the yield of any steps;

▼   change in the cost of a raw material from a supplier;

▼   changes in the price of energy such as electricity or gas;

▼   changing a solvent; and

▼   changing the quantities used.

The spreadsheet also breaks down each item's contribution to the final cost of the product. So if an item contributes only a few percent to the final cost, there is little point in striving to make savings on that item. However, if an item contributes 40% to the final cost then even a small saving can lead to a significant reduction in the final cost. *Figure 9* shows a typical spreadsheet. It uses non-systematic names as is usual in industry: Aniline is phenylamine. Lithal is lithium tetrahydridoaluminate (III). Ethylacetate is ethyl ethanoate. Toluene is methylbenzene.

Summary of the material costs and requirements for 100 000 kg of final materials:

Chemical	Step	Cost $/kg	kg Needed	Cost $	% of Raw Material	% of Total Cost
10% Lithal in THF	3	25.90	349.36	9048.38	73.60	25.33
43% Sodium hydroxide	3	0.19	132.65	25.20	0.21	0.07
Aluminium chloride	2	1.38	245.05	338.16	2.75	0.95
Aniline	1	2.50	84.39	210.87	1.72	0.59
Bromoisobutyryl bromide	1	10.64	109.79	1072.36	8.72	3.00
Celite	3	2.21	21.83	48.26	0.39	0.14
Conc. hychochloric acid	1	0.20	41.47	8.29	0.07	0.02
	2		336.73	67.35	0.55	0.19
Ethyl acetate	2	0.85	496.53	416.95	3.39	1.17
Water	1	0.00	397.20	0.00	0.00	0.00
	2	0.00	828.79	0.00	0.00	0.00
	3		61.14	0.00	0.00	0.00
Total water			1287.13	0.00	0.00	0.00
Toluene	1	0.30	688.74	266.62	1.65	0.58
	2		942.99	282.90	2.30	0.79
	3		1893.08	567.93	4.62	1.59
Total raw materials				12293.36	100.00	34.42

Notes
1) The 43% sodium hydroxide is used to quench the lithal, *ie* destroy the excess remaining after the reaction.
2) Celite is a powder used to aid filtration.
3) Concentrated hydrochloric acid is used in step 1 to wash the reaction mixture and dissolve any remaining aniline (which is basic). In step 2 it is used to dissolve excess aluminium chloride into an aqueous layer.
4) Ethyl acetate is used as a solvent for extracting the product of step 2.
5) Water is needed in step 3 to dilute the sodium hydroxide solution (which is sold as a 43% solution), in step 2 as a wash and in step 1 to dissolve unreacted aniline (as anilinium ions).
6) Toluene is needed as a solvent in the work up after stage 2.

Figure 9 A typical spreadsheet for the synthesis of 3,3-dimethylindoline (part of the BRL 46470A, supply route) in three steps

This spreadsheet refers to the production of 3,3-dimethylindoline in three steps by the supply route (*Fig 7*). Notice that materials comprise 34.42% of the total cost – the rest is process time (*ie* the cost of the plant and its operation) costed at $250 per hour). The most significant chemical cost is 10% lithal in THF (lithium tetrahydridoaluminate(III) in the ether-like solvent tetrahydrofuran) which represents 25.33% of the cost. It might be possible to make significant savings if it were possible to use less of this, buy it more cheaply or change to a cheaper reducing agent. In

RS•C

contrast, the cost of sodium hydroxide is negligible (a mere 0.07%) and it would be pointless to try to make savings here.

Over time, any synthetic method will be gradually improved, leading to reduced cost, although the scope for this gets less and less until the only possibility for saving becomes a change to a better synthetic method. This has been seen in *Fig 4* which shows improvements in the discovery, supply and manufacturing routes over time. It also illustrates the overlap between the development of these methods so that there is no interruption in the supply of the drug.

RS•C

Answers to questions

1. All three have a benzene-type ring fused to a five-membered ring containing at least one nitrogen atom. All three have a carbonyl functional group. All have a tertiary amine functional group.

   Teacher's note. The activity of all three drugs depends crucially on the distance between the nitrogen of the tertiary amine group (top right of each structure as drawn in *Fig 2*) and the oxygen of the carbonyl group. This distance must be seven atoms (as it is in all three drugs in *Fig 2* for them to act as $5HT_3$ antagonists.

   Serotonin, too, has a benzene-type ring fused to a five-membered ring containing a nitrogen atom. It has no carbonyl group. It has a secondary amine and a primary amine group but no tertiary amine.

2. N-H and O have the same number of electrons (they are isoelectronic). O has a lone pair instead of the shared pair in the N-H bond.

3. The tertiary amine group of 3,3-dimethylindoline will be basic and therefore react with the acid to form a salt. Base must be added to neutralise the acid and produce free 3,3-dimethylindoline before it can be crystallised.

4. (i)   The hydride ion, H⁻ is a nucleophile which attacks the $C^{\delta+}$ of the C=O group but not the electron-rich aromatic ring.

   (ii) The aromatic ring loses its aromatic stability if it is hydrogenated so that this is energetically unfavourable.

5. Bromoisobutyryl bromide $2.44/mole and phenylamine $0.23/mole

6. 2-bromo-2-methylpropanoyl bromide.